Cybersecurity Fundamentals for Beginners

Your Comprehensive Guide to Understanding Threats, Building Defenses, and Starting Your Career

Alton Hardin

ALTON TEACHES LLC

Editing by RJ Frampton

Cover Design by Aaftab Sheikh

ISBN-13: 978-0-9982945-3-7

About The Author

ALTON HARDIN is a former college professor, cybersecurity specialist, and the founder of Alton Teaches LLC. His journey in IT started when he was young, building computers for his family and helping out with his father's business. That early passion for technology led to over twenty years of working in information technology.

Alton is passionate about teaching. Over the years, he's had the privilege of helping more than 250,000 students around the world improve their IT and cybersecurity skills through his online courses. His goal has always been to make complex technical topics easy to understand, and he's proud to have been recognized as a top-rated IT instructor with seven best-selling courses on Udemy.

Alton started his professional IT career in 1999. Since then, he's worked in a variety of roles, including web development, desktop and network support, systems administration, SQL development, and database administration, before moving into cybersecurity. This diverse background has given him a well-rounded understanding of how cybersecurity fits into the bigger picture of technology.

In 2009, he made the switch to cybersecurity and spent the next 12 years as a subject matter expert in governance, risk, and compliance (GRC). Most of that experience came from his time as a cybersecurity specialist at the United States Department of the Treasury. Working in federal cybersecurity helped him develop the ability to break down complex concepts for beginners. Since he's worked in just about every IT role, from help desk support to security specialist, he has a good sense of how all the different pieces of technology fit together.

Alton holds two graduate degrees: an MBA in Information Assurance and Security Management and an MS in Information Technology Network Management. Over the years, he's also earned several IT certifications, including the Certified Information Systems Security Professional (CISSP #740196), ITIL, Security+, Network+, Project+, and Junior Penetration Tester (eJPT).

His love for teaching started early. He began as an undergraduate group tutor for chemistry students, and later, as a graduate student, he taught a GMAT prep course. After finishing his MBA, he became an adjunct professor at his local state university, where he taught Information Systems and IT Management courses.

Alton left his adjunct professorship when he moved to a new city. In 2016, he started his own business so he could keep sharing his passion for teaching online. After more than twenty years in IT and cybersecurity, he decided in 2022 to run his company full-time and focus on making technical education accessible to everyone.

Outside of IT, Alton is also a #1 Amazon best-selling author of poker books and the founder of MicroGrinder Poker School, where he offers affordable poker training for beginners and micro stakes players. His poker books have sold over 90,000 copies worldwide.

To learn more about Alton's IT courses and teaching philosophy, visit **InstructorAlton.com**. For poker learning resources, visit **MicroGrinder.com**.

Preface

My Passion for Teaching

Teaching has always been a passion of mine. From tutoring chemistry students as an undergraduate to teaching GMAT prep courses in graduate school, I've consistently found joy in helping others learn and succeed. After earning my MBA in Information Assurance and Security Management, I had the privilege of teaching as an adjunct professor at my local state university, where I taught Information Systems and IT Management courses to undergraduate students. That experience solidified my love for education and my desire to make complex technical topics accessible to everyone.

When I moved and could no longer teach at the university, I transitioned to online education and founded Alton Teaches LLC. This shift allowed me to reach a much broader audience and pursue my passion for teaching full-time. Over the past seven years, I've had the honor of teaching over 200,000 students worldwide through my IT and cybersecurity courses, with 7 of them becoming best-sellers. The positive feedback and success stories from my students continue to motivate me every day.

Why I Wrote This Book

My motivation for writing *Cybersecurity Fundamentals for Beginners* stems from my passion for teaching, my love for writing, and my desire to help address a critical need in the cybersecurity industry.

After publishing four successful poker books, including a perennial best-seller that's sold nearly 90,000 copies worldwide, I knew I could apply the same clear, approachable writing style to cybersecurity. Writing is a natural extension of my teaching—it allows me to reach even more people and provide them with a resource they can reference again and again. Following the success of my poker books, I felt confident that I could apply the same clear, approachable writing style to cybersecurity and create something truly valuable for aspiring IT professionals and curious learners alike.

At the same time, I observed a significant shortage of qualified cybersecurity professionals and a need for more truly beginner-friendly resources. In 2021 and 2022, there was a global shortage of over 3 million qualified cybersecurity professionals: positions that organizations desperately needed to fill but couldn't because knowledgeable candidates simply didn't exist. Many people who were interested in the field didn't know where to start, and many introductory cybersecurity resources were too technical for true beginners or failed to provide a comprehensive overview of the field.

Drawing from my 12 years of experience as a cybersecurity specialist at the United States Department of the Treasury, where I worked as a governance, risk, and compliance subject matter expert, I wanted to create a resource that would bridge this gap. I wanted to write a book that would serve as a gentle introduction to cybersecurity—one that would teach the fundamentals to aspiring IT professionals and help business professionals and everyday individuals better understand the cybersecurity landscape.

This book is the result of that vision. It's designed to be an accessible entry point into the world of cybersecurity, whether you're considering a career in the field, you're a business professional who needs to grasp cybersecurity concepts, or you're simply someone who wants to better protect yourself and your organization in an increasingly connected world.

What Makes This Book Different

This book takes a 10,000-foot perspective on cybersecurity, providing you with a comprehensive overview of essential concepts without getting you lost in overly technical details. I've structured the content to build progressively, starting with foundational concepts and gradually introducing more complex topics. My teaching approach, which I call the KISS method (Keep It Simple Student), ensures that even if you have no technical background, you'll be able to keep up with the material. Throughout my career in IT and cybersecurity, and through my years of teaching, I've developed an ability to break down complex technical concepts into simple, easy-to-digest explanations. That's exactly what I've done in this book. I assume you're new to cybersecurity and explain everything in plain language, using real-world examples and case studies to illustrate key concepts.

Complementary Course Access

This book also includes complementary access to my 4-hour *Cybersecurity for Absolute Beginners: Essential Concepts* online course, which normally sells for $49. The online course features video lectures, live demonstrations, and additional resources that complement the material in this book. To gain access to the online course, you'll need to provide me with proof of purchase, following the steps below:

1. Take a screenshot or picture of your receipt as proof of purchase
2. Email your proof of purchase to alton@instructoralton.com
3. I'll email you your free enrollment code

My Hope for You

Writing this book has been a labor of love, driven by my passion for teaching and my commitment to helping others succeed in the fields of information technology and cybersecurity. I've poured my knowledge, experience, and teaching expertise into creating this resource that makes cybersecurity accessible for beginners.

I sincerely hope you find this book valuable and that it helps you develop a strong foundational understanding of cybersecurity. Whether you're exploring a potential career path, seeking to better protect your organization, or simply wanting to explore the cybersecurity landscape, I'm here to guide you every step of the way. Remember, everyone starts somewhere. With dedication and the right resources, anyone can develop a solid cybersecurity foundation.

Thank you for choosing this book as your introduction to cybersecurity. I'm excited to be your guide on this journey.

— Alton Hardin

Contents at a Glance

Table of Contents

Introduction

Welcome to Cybersecurity

Welcome to *Cybersecurity Fundamentals for Beginners*. I'm glad you're here. Whether you're considering a career in the field, you're a business professional who needs to understand cybersecurity concepts, or you're simply curious about how to better protect yourself in our increasingly connected world, this book will provide you with a solid foundation.

Just about everything you do today is connected to the Internet. Most of us shop online regularly, browse the internet daily for news and information, and rely on online technologies for work, education, entertainment, and connecting with friends. Our homes are filled with connected devices, our cars are essentially computers on wheels, and our personal and professional lives are increasingly digital. This connectivity brings tremendous benefits, but it also creates vulnerabilities that cybercriminals are eager to exploit.

The threat is real and growing. The global financial cost of cyberattacks rose from $6 trillion in 2021 to $10.5 trillion in 2025. And if you thought hackers only targeted large businesses, you'd be wrong: they target everyone. From individuals to small businesses to Fortune 500 companies and government agencies, no one is immune to cyber threats.

This is why we need qualified cybersecurity professionals now more than ever. It's also why everyone, regardless of their profession, needs a basic understanding of cybersecurity. This book is designed to provide exactly that: a comprehensive introduction to cybersecurity fundamentals that anyone can understand.

What This Book Is

This book is a beginner-friendly introduction to cybersecurity: an accessible, comprehensive overview of essential IT security concepts from a 10,000-foot perspective. It's designed for people who are new to the field and need a broad understanding of cybersecurity principles, threats, and countermeasures.

Think of this book as your guided tour through the cybersecurity landscape. We'll cover a wide range of topics and material, spending time on many key points so you can develop a thorough understanding. The book provides a comprehensive overview of essential cybersecurity concepts and principles, an entry-level introduction to IT security, presented in an easy-to-understand manner.

Throughout the book, I'll explain cybersecurity terms in plain language, doing my best to avoid unnecessary jargon and technical complexity. My goal is for you, by the end of the last chapter, to have a strong foundational understanding of cybersecurity that'll serve you well, whether you're pursuing a career in the field or simply want to be more informed about cybersecurity in your personal and professional life.

What This Book Isn't

To set expectations, let's clarify what this book is not:

This Is Not a Hands-On Technical Training Guide: While I'll explain various cybersecurity concepts and demonstrate some ideas, this isn't a step-by-step, how-to guide that will teach you to configure firewalls, set up intrusion detection systems, or perform penetration testing. Those topics require hands-on technical training that goes beyond the scope of an introductory book.

This Is Not an Ethical Hacking or Penetration Testing Guide: While I'll teach you about the methodologies involved in ethical hacking and penetration testing, and I'll explain how hackers think and operate, this isn't a book that will teach you to become a penetration tester or ethical hacker. I will, however, provide resources for further learning if you're interested in pursuing those paths.

This Is Not a Certification Study Guide: While the knowledge you gain will certainly help if you decide whether to pursue cybersecurity certifications like the CompTIA Security+ or others, this book isn't specifically designed to prepare you for any particular certification exam. It's designed to give you a broad, foundational understanding of cybersecurity concepts.

This Is Not an All-Encompassing Cybersecurity Encyclopedia: Cybersecurity is a vast and constantly evolving field. This book covers essential concepts and fundamentals, but it doesn't attempt to cover every possible topic, tool, or technique in cybersecurity. Instead, it focuses on giving you a solid foundation that you can build upon as you continue your learning journey.

How to Use This Book

This book is designed to make cybersecurity accessible to absolute beginners. To help you navigate the content and get the most out of your learning experience, I've incorporated several special features throughout the book.

Callout Boxes: Your Reading Guide

Throughout this book, you'll encounter special boxes and icons that help you identify different types of content and prioritize what matters most for your learning journey:

🔍 **DEFINITION**: This box highlights important terms and their meanings. Use it to quickly understand key vocabulary as you read.

🎯 **KEY POINT**: Critical takeaways you should remember. These are foundational elements that everything else builds upon.

💡 **EXAMPLE**: Real-world scenarios that illustrate how concepts apply in practice, helping you connect theory to reality.

⚠️ **COMMON MISTAKE**: Pitfalls to avoid. Learning from others' mistakes is faster than making them yourself.

📖 **TECH TALK**: Optional technical details for readers who want to go deeper into the technology. If you're just starting out, you can skip these and come back later.

📰 **GOING DEEPER**: Optional additional context that provides more background or explores related topics. These callouts enrich your understanding but aren't essential for grasping the core principles.

🔧 Quick Actions: Learning by Doing

Each chapter concludes with a Quick Action, a practical exercise you can complete to immediately apply what you've learned. Reading about cybersecurity builds understanding, but learning by doing is how you truly master it.

What Makes Quick Actions Different

Quick Actions are designed to be immediately actionable, progressive (each builds on previous actions), and personal (focused on your home devices and accounts, not work systems).

How to Approach Quick Actions

You don't need to complete every Quick Action immediately. However, I encourage you to at least read each one as you finish its chapter, mark the ones most relevant to your situation, and complete them within a week of finishing the book. By completing all the Quick Actions, you'll dramatically improve your personal cybersecurity posture while also gaining practical experience.

Case Studies

Throughout the book, you'll find detailed case studies examining real-world cybersecurity failures at major organizations. These case studies bring the concepts to life by showing you exactly what happens when security fails and what you can learn from those failures. These aren't just stories; they're learning opportunities that demonstrate why the principles and practices we discuss matter in the real world.

Book Overview & Structure

This book is structured in a logical, progressive manner, with each chapter building on the concepts learned in the previous ones. The book follows a five-part structure designed to take you from foundational knowledge through threats to comprehensive defenses, and finally to your path forward in cybersecurity.

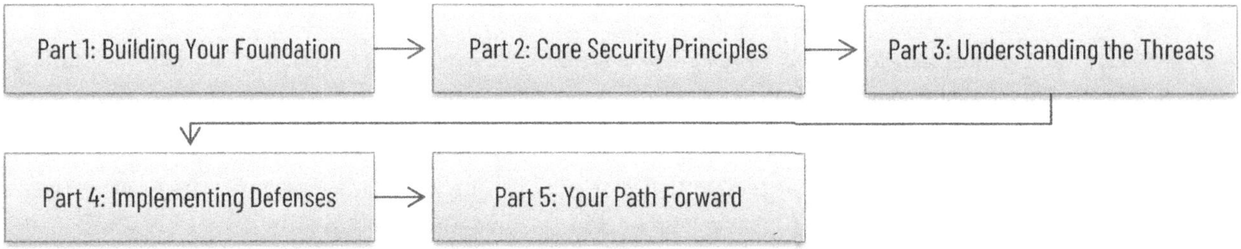

Part 1: Building Your Foundation (Chapters 1–3)

Chapter 1: Introduction to Cybersecurity: We'll start by exploring what cybersecurity is, why it matters, the current state of the industry, and the various career paths in the field.

Chapter 2: Understanding Hackers: You'll learn about the different types of hackers, their motivations, and the methodologies they use to compromise systems and networks.

Chapter 3: Networking 101: Since much of cybersecurity is rooted in networking, we'll take a crash course in computer networking fundamentals to ensure you understand the basics.

Part 2: Core Security Principles (Chapter 4)

Chapter 4: Core Cybersecurity Principles: Before diving into specific threats, we'll establish the fundamental defensive principles that guide cybersecurity practice: the CIA and DAD triads, defense in depth, least privilege, AAA (authentication, authorization, and accounting), non-repudiation, and risk management basics. These concepts form the lens through which you'll evaluate everything that follows.

Part 3: Understanding the Threats (Chapters 5–7)

Chapter 5: Malware Threats: We'll explore various types of malicious software, including viruses, worms, trojan horses, ransomware, spyware, rootkits, and emerging threats like AI-enhanced malware.

Chapter 6: Additional Threats & Vulnerabilities: Beyond malware, we'll examine threats such as zero-day attacks, denial-of-service attacks, protocol spoofing, web-based attacks, insider threats, and AI-enabled attacks.

Chapter 7: Social Engineering & Email-Based Threats: You'll learn about the human element of cybersecurity, including phishing, spear phishing, whaling, business email compromise, and other social engineering tactics that manipulate people rather than exploit technology.

Part 4: Implementing Defenses (Chapters 8–11)

Chapter 8: Fundamental Threat Countermeasures: You'll learn about the specific tools and techniques used to protect against cyber threats, including multi-factor authentication, password management, endpoint protection, encryption, firewalls, intrusion detection systems, VPNs, and wireless security.

Chapter 9: Security Operations: We'll explore how organizations actively defend themselves through continuous monitoring, endpoint detection and response, incident response processes, security assessments, and red team versus blue team exercises.

Chapter 10: Security Management & Culture: You'll learn about the organizational and governance aspects of cybersecurity, including legal compliance, patch management, change management, formal risk management programs, personnel policies, and security awareness training.

Chapter 11: Application Security, Resilience, & Modern Devices: We'll explore how to build secure applications from the ground up using the Secure Software Development Lifecycle (SDLC), examine cyber-resiliency strategies that ensure business continuity when attacks succeed, and address the unique security challenges of modern technologies, including BYOD policies, IoT devices, and virtualized environments.

Part 5: Your Path Forward (Chapters 12 & 13)

Chapter 12: Getting Started in Cybersecurity: We'll discuss how to break into the cybersecurity field, including entry-level roles, foundational certifications, practical experience opportunities, and building your professional network.

Chapter 13: Conclusion & Congratulations: We'll wrap up your journey through cybersecurity fundamentals, celebrate your achievement, and provide final guidance as you continue your cybersecurity education and career.

Throughout the book, you'll encounter detailed case studies examining real-world cybersecurity failures at major organizations like Equifax (Chapter 1), Sony (Chapter 6), and Target (Chapter 9). These case studies demonstrate why the principles and practices we discuss matter in the real world.

My Teaching Approach: KISS Method

This book follows a KISS (Keep It Simple Student) approach to teaching cybersecurity. Throughout the book, I assume the concepts presented are new to you and teach them in an easy-to-understand manner. I've designed the book to teach progressively, beginning with the basics, building your foundational knowledge, and then introducing more complex principles and topics.

I strongly recommend you read each chapter sequentially, from first to last, rather than jumping around from chapter to chapter. Each chapter builds on the knowledge from previous chapters, and the book is specifically designed to be read in order.

When you encounter technical terms you don't understand, you can refer to the comprehensive glossary at the end of the book for quick definitions. I've also included diagrams, illustrations, and real-world examples throughout to help clarify the teaching points and make the material more engaging and memorable.

Why You Should Read This Book

This book is for anyone who wants to understand cybersecurity fundamentals:

- Aspiring IT professionals considering a career in cybersecurity
- Business professionals and managers who need to understand cybersecurity basics to make informed decisions
- Students exploring potential career paths in technology
- Everyday people who want to better protect themselves and their families online
- Career changers looking to transition into the growing field of cybersecurity

You don't need any prior technical knowledge or IT experience to benefit from this book. If you can use a computer and browse the Internet, you have all the background knowledge you need to get started.

By the end of this book, you'll have a comprehensive understanding of cybersecurity fundamentals. You'll understand the threat landscape, the principles that guide cybersecurity practice, the tools and techniques used to protect systems and data, and how organizations approach cybersecurity. Most importantly, you'll have a launchpad for continuing your learning journey, whether that means pursuing a career in cybersecurity or simply becoming more security-conscious in your daily life.

I'm excited to have you here and look forward to guiding you through the essential concepts of cybersecurity. Let's get started!

PART 1: BUILDING YOUR FOUNDATION

Overview

Before you can protect systems and networks, you need to understand what you're protecting and who you're protecting against. This opening section is your crucial starting line, establishing the essential foundation for everything that follows in this book.

In Chapter 1, you'll discover why cybersecurity matters in today's interconnected world and explore the diverse career opportunities available in this rapidly growing field. Chapter 2 takes you inside the mindset of hackers—both the malicious attackers and the ethical security professionals who think like them—revealing the systematic methodology used to compromise systems. Finally, Chapter 3 provides a crash course in networking fundamentals, which provides the technical vocabulary and concepts you'll need to truly understand how attacks unfold and how defenses work.

Think of this section as learning the rules of the game before you start playing. By the end of Part 1, you'll understand the "why" behind cybersecurity, the "who" you're defending against, and the "what"—the networks and systems that form the battlefield where security professionals and attackers meet.

Chapter 1. Introduction to Cybersecurity

Introduction

This chapter introduces you to the world of cybersecurity: what it is, why it matters, and why you need to understand it in today's interconnected world. You might be wondering why you should care about cybersecurity, especially if you're not planning to work in IT. The reality is that cybersecurity affects everyone. Whether you're checking your email, shopping online, using social media, or simply browsing the web, you're interacting with systems that need to be protected from cyber threats. Knowing the basics of cybersecurity isn't just for IT professionals. It's essential knowledge for anyone living in the digital age.

This chapter will help you understand why cybersecurity is so critical to our modern way of life and what's at stake when security fails. You'll get an overview of the diverse field of cybersecurity and its various specializations. Whether you're considering a career in cybersecurity or simply want to learn about this essential aspect of our digital world, this foundation will serve you well. For those of you considering a career in cybersecurity, we'll explore career paths, certifications, and how to break into the field in much greater detail in Chapter 12.

To understand just how high the stakes are, we can look at one of the most significant data breaches in history—one that affected nearly half of all Americans and serves as a stark reminder of why cybersecurity matters.

Case Study: Equifax Web App Vulnerability

Background

In September 2017, the credit reporting agency Equifax announced a data breach of staggering scale. The private information of 147 million Americans, nearly half the country's population, had been stolen by hackers. They stole names, Social Security numbers, birth dates, addresses, driver's license numbers, and over 200,000 credit card numbers. The attack lasted from May through July 2017 before being discovered.

The Attack

The hackers exploited a vulnerability in the software that Equifax used for its website. This vulnerability had a known fix, a patch released in March 2017, 2 months before the attack began. However, Equifax didn't install this patch on all its systems.

The attackers exploited this unpatched vulnerability to break into Equifax's servers. Once inside, they moved throughout the network, found databases containing customer information, and stole data for 76 days without being detected. Equifax's security systems failed to catch the breach because the IT staff had let the security certificates expire. These certificates are what allow security tools to inspect encrypted traffic. Without valid certificates, the security tools couldn't see what data was leaving the network.

Impact

Equifax paid approximately $700 million to settle with the Federal Trade Commission and state governments, including $380 million for a victim compensation fund. The company spent hundreds of millions more on security improvements and on providing credit monitoring to affected customers.

The CEO resigned and Equifax's stock price fell by more than 30%. Congressional hearings revealed the company's security failures to the public, severely damaging its reputation.

For the victims, the consequences were long-lasting. Social Security numbers can't be changed, leaving millions vulnerable to identity theft for years to come. Unfortunately, because of this, many people spent years dealing with fraudulent accounts and damaged credit scores.

Lessons Learned

- **Install Security Patches Promptly**: A patch existed months before the breach, but Equifax failed to apply it to all systems. Organizations should quickly identify, test, and install security updates across their entire environment.
- **Use Multiple Layers of Security**: Multiple security failures enabled the breach: unpatched software, expired certificates, poor network segmentation, and inadequate monitoring. Layered security could have stopped or limited the attack even after the initial breach.
- **Leadership Must Prioritize Security**: Equifax was aware of the vulnerability but failed to prioritize fixing it. Security should be treated as a critical business function with clear accountability at the leadership level.

What Is Cybersecurity?

Cybersecurity is the practice of protecting systems, networks, devices, and the data that resides on them, as well as the programs that these systems use. The goal is to protect all of this from unauthorized access and digital attacks, i.e., cyberattacks.

> **DEFINITION: Cybersecurity** is the practice of protecting IT systems, networks, data, and programs from unauthorized access and digital attacks.

When it comes to cybersecurity, you may also hear people refer to it as IT security or data privacy. All three are interrelated. We have information security (or IT security), privacy, and cybersecurity. Typically, cybersecurity falls under the larger umbrella of information security and data privacy. Different organizations may have alternative definitions and scopes for these terms. One organization may have a cybersecurity department, while another may call the same section its information security department. The important thing to understand is that they're all intertwined.

As we progress through this book, some topics will relate to privacy or information security beyond just cybersecurity, as they all play a significant role in protecting organizational assets, primarily digital assets.

You Don't Need to Be a Tech Wizard

If you're reading this book, you might be wondering whether you need to be a tech wizard or have an advanced technical background to understand cybersecurity. The answer is no. While cybersecurity involves technology, it's fundamentally about protecting people, data, and organizations. You don't need to be a coding expert to work in this field or to understand how to protect yourself and your organization from cyber threats.

Many cybersecurity roles focus on policy, compliance, risk management, training, and communication—skills that don't require coding expertise. Even technical roles like security analysts often rely more on critical thinking, problem-solving, and understanding how systems work rather than being a tech wizard. This book is designed for absolute beginners, and we'll explain technical concepts in plain language. If you encounter a term you don't know, remember that we've included a comprehensive glossary in the back of the book for quick reference.

Why Do We Need Cybersecurity?

Everything Is Connected

Think about your daily life for a moment. Your smartphone wakes you up in the morning. Your smart thermostat adjusts the temperature in your house. Your car connects to your phone for navigation and music. Your TV streams your favorite shows. Your doorbell camera alerts you to visitors. Even your refrigerator might send you notifications when you're running low on groceries.

This level of connectivity would have seemed like science fiction just a couple of decades ago, but it's now our everyday reality. We've moved far beyond just computers and smartphones being connected to the Internet. Today, we live in the era of the Internet of Things (IoT), where virtually every device in our homes, cars, and workplaces is always connected to a network.

The convenience is undeniable. You can adjust your home's temperature from across the country, unlock your door for a delivery person while you're at work, or check your security cameras from anywhere in the world. But here's the critical point: If you can access these devices remotely, so can someone else (if they know how to get in).

Every connected device represents a potential entry point for cybercriminals. When your refrigerator, thermostat, doorbell, and car are all connected to the Internet, each one becomes a possible target. And unlike in the past, when a burglar had to physically break into your home, today's threats can come from anywhere in the world, at any time, through any one of your connected devices.

This is why cybersecurity matters. It's not just about protecting computers anymore; it's about protecting every aspect of your increasingly digital lives.

DEFINITION: Internet of Things (IoT) refers to everyday devices that connect to the Internet: smart thermostats, fitness trackers, security cameras, and more. While convenient, each connected device is a potential entry point for attackers.

Protecting Our Organizations

Cybersecurity isn't just a personal concern; it's essential for every organization, whether you're running a business, nonprofit, or government agency. Organizations have a responsibility to protect their assets: IT infrastructure, customer and employee data, intellectual property, and more. This isn't just good practice; it's both a legal obligation and a strategic necessity.

The Trust Factor

When organizations fail to protect their data and systems, they lose something invaluable: trust. Customers, partners, and stakeholders expect their information to be secure. A data breach doesn't just expose information—it shatters confidence. We've seen it happen time and time again, where major retailers and corporations suffer breaches, customer data is stolen, and relationships between the businesses and customers are compromised. Customers move to competitors, business partners become wary, and the damage inevitably extends far beyond the immediate financial costs.

Beyond Business: Critical Infrastructure

The stakes also extend beyond individual organizations. Cybersecurity is critical to national security and public safety. Power grids, water treatment facilities, transportation systems, and healthcare networks all depend on a secure digital infrastructure. A successful attack on any one of these systems could have devastating consequences for entire communities or nations, not just individual companies.

Productivity & Business Focus

Effective cybersecurity also drives productivity. When organizations lack adequate security measures, they spend their time "putting out fires," constantly responding to threats, incidents, and breaches. This reactive approach drains resources and diverts attention from core business objectives. Instead of growing the business and serving customers, teams are stuck dealing with security crises that could have been prevented.

Organizations that invest in strong cybersecurity don't just avoid losses; they create a competitive advantage. They're able to focus on innovation and growth rather than damage control. They maintain customer trust, protect their reputation, and demonstrate to their partners, regulators, and customers that they take their responsibilities seriously.

The State of Cybersecurity

Now that you understand what cybersecurity is and why we need it, you might be wondering: just how bad is the threat? The threat is very real and growing rapidly. However, these aren't just abstract statistics. Behind every number is a real person whose identity was stolen, a small business that closed its doors, or a hospital that couldn't access patient records.

As I write this book in 2025, I'm looking back at data from previous years, the same approach I take with my students each year to understand where threats are heading. The picture is clear: threats are becoming more frequent, more sophisticated, and more costly. Nevertheless, opportunities to make a difference in this field are greater than ever.

Understanding the Trends (2021–2024)

The Ransomware Surge

Between 2021 and 2022, ransomware attacks grew by over 350% annually. Furthermore, by 2024, they surged another 84%, accounting for 35% of all cyberattacks. Why? It's profitable (attackers demand millions in cryptocurrency), it's easy (Ransomware-as-a-Service allows even non-technical criminals to launch attacks), and it works (many organizations pay rather than face extended downtime).

> **EXAMPLE:** In May 2021, a ransomware attack forced the Colonial Pipeline, which supplies 45% of the fuel consumed on the East Coast, to shut down operations, causing fuel shortages across multiple states. The company paid $4.4 million in ransom to restore access to their systems. This incident showed how cyberattacks can have real-world consequences far beyond stolen data.

Small Businesses Under Attack

Here's a dangerous myth: "We're too small to be a target. Hackers only go after big companies."

In 2021 and 2022, 43% of all cyberattacks targeted small businesses, yet only 14% of these businesses felt prepared to defend themselves. Hackers choose them because they're easier targets with fewer security resources, and they serve as stepping stones to larger organizations through vendor relationships.

> ! **COMMON MISTAKE:** Believing "we're too small to be targeted" is one of the most dangerous assumptions a small business can make. Attackers look for vulnerabilities, not company size.

The AI Arms Race

By 2024, phishing incidents had surged by 1,265%, over 12 times their previous levels. Phishing is when attackers send fraudulent messages that appear to be from legitimate sources to trick you into revealing passwords or clicking on malicious links. Generative AI has also enabled attackers to create highly personalized phishing emails that are much harder to detect, making these attacks far more convincing than ever before. Additionally, deepfake technology has enabled attackers to create convincing audio or video of executives, tricking employees into transferring money or revealing sensitive information. We'll explore phishing in Chapter 7.

Expanding Attack Surfaces

Cloud-based attacks increased by 75% in 2024 as more organizations migrated their data to cloud platforms. In addition, IoT breaches hit $1.51 billion in just the first 6 months of 2022, a 77% increase from 2021. Moreover, supply chain attacks affected over 1.83 million users and businesses in 2024, a 33% year-over-year increase.

The Real-World Impact

The Financial Impact

Global cybercrime costs grew from $6 trillion in 2021 to $8 trillion in 2023, projected to hit $10.5 trillion by 2025. If cybercrime were a country, it would have the world's third-largest economy. Cybercrime isn't just an IT problem—it's an economic issue that affects everyone. These costs are passed on through higher prices, increased insurance premiums, and disrupted services. Additionally, the average data breach cost jumped to $4.88 million in 2024, with healthcare organizations facing the highest costs at $9.77 million per breach.

The Detection Problem

Organizations took an average of 258 days to detect and contain breaches in 2024, with credential-based breaches extending to 328 days (nearly a full year). The longer attackers remain undetected, the more damage they can do: stealing more data, establishing multiple backdoors, and moving throughout networks. Consequently, every day of delayed detection increases the cost and scope of the breach.

What's Being Stolen

Cybercriminals stole over 33 billion records in 2023, a 175% increase from 2018. Furthermore, by 2024, 46% of breaches involved personally identifiable information, such as Social Security numbers, tax IDs (or ITINs), and other sensitive information. Unlike a stolen credit card, these identifiers can't be changed, leaving victims vulnerable to identity theft for years or decades.

The Workforce Crisis: Opportunity in Disguise

The cybersecurity workforce shortage grew from 3.12 million professionals in 2021 to 4.8 million by 2024. Moreover, 53% of organizations faced critical staffing shortages, resulting in $1.76 million in additional breach costs and 6 to 12 months of delays in filling positions.

However, this workforce shortage represents a genuine opportunity. Organizations are actively seeking people who want to learn cybersecurity skills. They're willing to train candidates who show aptitude, even without extensive technical backgrounds. Organizations need people who understand how to protect digital assets, and we'll explore how to break into this field in Chapter 12.

What This Means For You

My goal in sharing these statistics isn't to scare you; it's to highlight the importance of being vigilant and informed. The threats are real and affect everyone, but most attacks are preventable through basic security practices. The Equifax breach happened because of an unpatched vulnerability. Many ransomware attacks succeed because of weak passwords or a lack of multi-factor authentication.

Understanding these threats is the first step toward protecting yourself and your organization. Additionally, there's a significant opportunity; people willing to learn cybersecurity skills are in high demand. You don't need to be an expert to make a difference. Simple practices, such as using strong passwords, enabling multi-factor authentication, and keeping software up to date, can significantly reduce your risk. The state of cybersecurity is serious, but it's not hopeless. Every person who learns these concepts and implements basic security practices makes the digital world safer. That includes you.

Cybersecurity Is Diverse

A Primer for Complete Beginners

Before we explore the different areas of cybersecurity, it helps to understand what makes this field so diverse and why that's actually good news for you as a beginner.

Think of cybersecurity as a large city with many neighborhoods, each with its own character and purpose. Just as a city needs police officers, firefighters, city planners, teachers, and healthcare workers to function, cybersecurity needs people with different skills and interests working together to protect our digital world.

You might be wondering: "Where do I fit in?" The beauty of cybersecurity's diversity is that there's likely a place for you, regardless of your background. Are you detail-oriented and enjoy solving puzzles? You might excel in threat analysis. Do you enjoy working with people and explaining complex ideas simply? Security awareness training or compliance might be your calling. Are you creative and think like an attacker? Penetration testing could be a great fit.

As you read through the following sections, don't worry if you're not familiar with every term or concept yet. We'll cover these topics throughout this book. For now, just get a sense of the breadth of the field and start thinking about what might interest you. Remember, cybersecurity isn't just one thing: it's a collection of interconnected disciplines, each essential to protecting our digital lives. As you learn more, you'll discover which areas resonate with you and where you can make your unique contribution.

Core Areas of Cybersecurity

Cybersecurity encompasses many specializations. We'll explore the core domains you're most likely to encounter as you begin your journey in this field.

Network Security focuses on protecting computer networks from intruders, whether they're targeted attackers or opportunistic malware. This involves monitoring network traffic, configuring firewalls, implementing intrusion detection systems, and ensuring that only authorized users can access network resources. Network security is often the first line of defense against external threats.

Application Security involves keeping software and devices free of threats. A compromised application could provide access to the data it's designed to protect, making this area critical. Application security professionals work to identify and fix software vulnerabilities, implement secure coding practices, and test applications for security flaws before deployment.

Cloud Security has emerged as one of the fastest-growing specializations as organizations increasingly move their data and applications to cloud platforms like AWS, Azure, and Google Cloud. Cloud security professionals implement controls to protect data, applications, and infrastructure in cloud environments, while navigating the shared responsibility model between cloud providers and organizations.

Identity and Access Management (IAM) focuses on ensuring that the right individuals have access to the right resources at the right times for the right reasons. This includes managing user identities, authentication methods (like passwords and multi-factor authentication), and authorization policies across an organization. In many ways, IAM is the foundation of modern security—if you can't control who has access to what, you can't protect anything.

Incident Response involves preparing for, detecting, analyzing, and recovering from security incidents. Incident response professionals are the first responders when a security breach occurs, working to contain the damage, investigate what happened, and restore normal operations. This domain also encompasses disaster recovery and business continuity planning, ensuring organizations can recover quickly from cyberattacks or other disruptions.

Governance, Risk, & Compliance (GRC) ensures that organizations meet regulatory requirements, manage cybersecurity risks effectively, and maintain appropriate policies and procedures. GRC professionals bridge the gap between technical security measures and business objectives, translating technical risks into business language that executives can understand and act on.

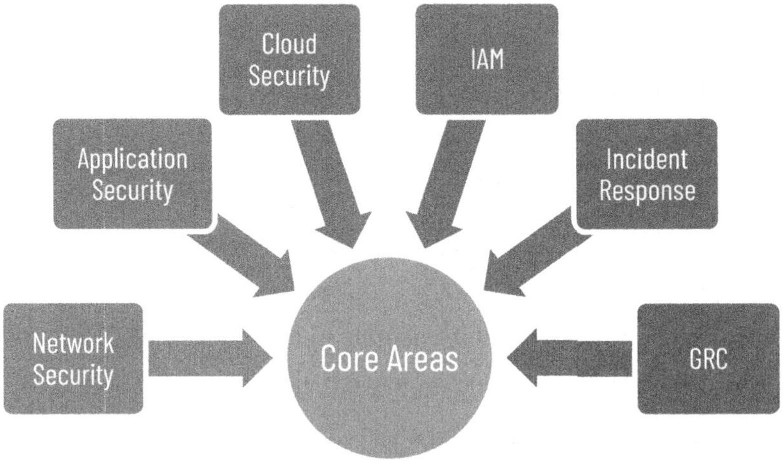

Figure 1.1: Core Areas of Cybersecurity

These six domains represent the foundation of modern cybersecurity. While many other specialized areas exist, such as forensics, threat intelligence, and operational security, understanding the core domains will give you a solid foundation as you continue your cybersecurity journey.

KEY POINT: You don't need to master all these areas. Most cybersecurity professionals specialize in one or two domains and collaborate with others who have complementary expertise. The key is finding an area that aligns with your interests and strengths.

Common Cybersecurity Roles

The diversity of cybersecurity extends to the many roles and career paths available in the field. Rather than overwhelming you with every possible position, let's focus on some roles you're most likely to encounter as you explore this field.

Security Analysts monitor networks and systems for security breaches, investigate violations, and implement security measures to protect an organization's computer systems and networks. This is one of the most common entry points into cybersecurity, offering broad exposure to various aspects of the field and providing an excellent foundation for future specialization.

Security Operations Center (SOC) Analysts monitor security alerts in real time, triage incidents, and coordinate responses to security events. SOC analysts work in shifts to provide 24/7 monitoring, to learn and recognize attack patterns, and to respond to threats as they emerge.

Security Engineers design, implement, and maintain security systems and infrastructure, including firewalls, intrusion detection systems, and encryption protocols. This role focuses on building and maintaining the technical controls that protect organizations from cyber threats.

Incident Responders respond to security breaches and cyberattacks, working to contain incidents, investigate root causes, and restore normal operations. These professionals are the first responders when a security breach occurs, often working under pressure to minimize damage and recover quickly.

Penetration Testers (Ethical Hackers) simulate cyberattacks to identify vulnerabilities in systems, networks, and applications before malicious hackers can exploit them. This role appeals to creative, curious individuals who think like attackers and enjoy the challenge of finding weaknesses in defenses.

Compliance Analysts ensure that organizations meet regulatory requirements and industry standards, conduct audits, and maintain documentation of compliance efforts. This role is ideal for detail-oriented individuals who enjoy working with policies, procedures, and frameworks.

Security Awareness Trainers educate employees about security threats and best practices, develop training programs, and promote a culture of security within organizations. This role leverages effective communication skills and is suitable for individuals who enjoy working with people and explaining complex ideas in a clear and simple manner.

These roles represent just a sample of the career opportunities available in cybersecurity. Each role requires specialized skills, appeals to different interests, and offers unique challenges and rewards. We'll explore several of these roles in much greater depth in Chapter 12, including entry-level positions, recommended certifications, and practical steps for breaking into the field.

Many successful cybersecurity professionals start in non-security IT roles (help desk, system administration, network administration) and transition into security. Others have come from completely different fields, such as teaching, military service, law enforcement, or even the liberal arts, and successfully transitioned into cybersecurity. What this means is that your path doesn't have to be linear, and your previous experience often provides unique perspectives that make you valuable in this field.

Bringing It All Together

Cybersecurity is the foundation of trust in our digital world. As you've learned in this chapter, it's not just about technology—it's about protecting people, organizations, and the critical systems we all depend on.

The Equifax breach demonstrated what happens when security fundamentals fail: a single unpatched vulnerability led to the exposure of personal data for 147 million people and cost the company approximately $700 million. This wasn't a sophisticated attack; in fact, it was preventable with basic security practices.

Understanding that everything, from our smartphones and smart homes to our cars and critical infrastructure, is now connected to the Internet helps explain why cybersecurity has become essential knowledge for everyone, not just IT professionals. Each connection represents both convenience and risk.

The field's diversity is one of its greatest strengths. Whether you're interested in technical roles like penetration testing, analytical work like security analysis, or people-focused positions in training and compliance, there's a place for you. The cybersecurity workforce shortage means opportunities exist at all levels for anyone willing to learn. As you continue through this book, remember that every concept builds on what you've learned here. The threats are real and evolving, but so are our defenses. Your journey into cybersecurity began with understanding why it matters. Now that you do, let's explore the first step in that defense: understanding the adversaries you'll be defending against.

Key Takeaways

- **Cybersecurity protects systems, networks, devices, and data** from unauthorized access and cyberattacks, and is closely related to information security and privacy.
- **Everything is connected to the Internet**, from personal devices to smart homes, cars, and critical infrastructure, creating an expanded attack surface that requires constant vigilance and protection.
- **Organizations need cybersecurity** to protect assets, maintain customer confidence, meet legal obligations, support national security, and increase productivity by preventing costly security incidents.
- **Current threats are sophisticated and pervasive**, with ransomware accounting for 35% of all cyberattacks, average breach costs exceeding $4.88 million, and organizations taking an average of 258 days to detect and contain breaches.
- **Cybersecurity is a diverse field** with numerous specializations and career paths for technical and non-technical professionals. You don't need to be a coder or engineer to contribute meaningfully to the field.

Chapter 2. Understanding Hackers

Introduction

Before you can effectively defend against cyber threats, you need to understand the adversaries you're facing. This chapter focuses on understanding hackers: who they are, what motivates them, and how they operate.

Not all hackers are the same. Some hack for financial gain, others for political or ideological reasons, and still others work to protect organizations from cyberattacks. There are also significant misconceptions about what hacking actually is, largely perpetuated by movies and television shows that dramatize the process.

In this chapter, we'll dispel those myths and give you a realistic understanding of how hackers operate. We'll explore the different types of hackers, ranging from black hat criminals to white hat security professionals, and examine what drives them, from script kiddies (a derogatory term for unskilled hackers) to sophisticated state-sponsored actors. Finally, we'll walk through the systematic methodology that both malicious and ethical hackers use to compromise systems and networks.

Understanding hackers isn't just academic. It's practical. When you understand how attackers think and operate, you're better equipped to defend against them. Whether you're an aspiring cybersecurity professional or simply someone who wants to protect themselves and their organization, this chapter will give you critical insights into the minds and methods of both malicious and ethical hackers.

What Hacking Isn't

Before we dive into the world of hackers and hacking, let's address something important: What you've seen in movies is almost certainly wrong.

We've all seen dramatic hacking scenes. A character sits at a computer, fingers flying across the keyboard at impossible speeds as lines of green code cascade down the screen. Within minutes, sometimes seconds, they've broken through multiple layers of security, cracked complex encryption, and gained access to classified government databases. Fancy graphics flash across the screen, progress bars fill up rapidly, and suddenly: "I'm in!" they announce triumphantly, often with authorities crashing through the door or while racing against a ticking clock. It looks exciting. It looks cool. And it's almost entirely fiction.

The reality of hacking differs dramatically from what Hollywood portrays. Ninety-five percent of the time, movies and TV shows glorify or exaggerate hacking scenes to make them more entertaining. That's understandable—watching someone actually hack would make for terrible television. Real hacking is often slow, methodical, and, frankly, quite boring to watch.

DEFINITION: Hacking is the act of gaining unauthorized access to computer systems, networks, or data. While the term can technically apply to any clever technical solution, in cybersecurity, it typically refers to exploiting vulnerabilities to bypass security measures.

For those of you interested in becoming ethical hackers, you need to understand that hacking is often a very slow and methodical process. It doesn't take minutes; it often takes hours or even days. When someone takes a hands-on ethical hacking certification exam, they're typically given anywhere from 24 hours up to several days to complete it. That's a significant amount of time to hack into systems and complete the required objectives, and that's just for a certification exam.

Real hacking involves extensive research, careful planning, patient observation, and systematic testing. A hacker might spend days or weeks gathering information about a target before even attempting to gain access. They might try hundreds or thousands of approaches, most of which fail. They document everything meticulously. They work through problems step by step, often hitting dead ends and having to backtrack. There's no dramatic countdown timer and no flashy graphics—just a person (or team) working methodically through a complex technical challenge, often staring at screens full of text, running automated tools, and waiting for results. So when you're watching hacking portrayed on TV or in the movies, remember: it's pure entertainment. The real world of hacking, both malicious and ethical, is far more complex, time-consuming, and technical than what you see on screen.

Types of Hackers

Now that we've cleared up what hacking isn't, it's time to talk about what hackers actually are. Not all hackers are the same, and understanding the different types is crucial to understanding cybersecurity.

Black Hat Hackers

Black hat hackers are what most people think of when they hear the word "hacker." These are hackers who violate computer security for malicious reasons: terrorism, espionage, or personal gain. They're the stereotypical bad guys of cybersecurity. They illegally hack into IT systems for a variety of malicious reasons, exploiting vulnerabilities without permission and often with criminal intent. Their goals typically include causing harm, stealing and ransoming data, and financially extorting individuals and organizations.

When you hear about a major data breach in the news, when a company's customer information is stolen, when ransomware locks up a hospital's systems, or when a website is defaced, these are the work of black hat hackers. They operate outside the law and their activities cause billions of dollars in damages every year. Black hat hackers may work alone, in small groups, or as part of larger criminal or governmental organizations. Some black hat hackers are highly skilled and sophisticated, while others simply use tools created by others. What they all have in common is that they're breaking the law and causing harm.

White Hat Hackers

On the opposite end of the spectrum, we have white hat hackers, who are the good hackers. White hat hackers, also known as ethical hackers, are cybersecurity professionals who use hacking techniques to identify and fix security vulnerabilities within legal boundaries. They apply the same techniques and tools that malicious hackers use to test and protect IT systems, but they do so legally, ethically, and with explicit permission from the organization to strengthen its defenses.

White hat hackers work in the cybersecurity divisions of companies and governmental organizations, for consulting firms, or as independent security researchers. They conduct vulnerability assessments, penetration tests, or pen tests (simulated cyberattacks designed to identify exploitable vulnerabilities), and other security assessments to enhance an organization's security posture. Their goal is to find vulnerabilities before the bad guys do, so those vulnerabilities can be fixed.

The key difference between white hat and black hat hackers isn't necessarily their technical skills; in fact, they're often equally adept at using many of the same tools and techniques. The difference is permission and intent. White hat hackers have explicit authorization to test systems; they work within defined boundaries, and their goal is to improve security rather than exploit it.

Gray Hat Hackers

There's also a third category that falls somewhere in between: gray hat hackers. These hackers operate in a moral and legal gray area. They might hack into systems without permission (which is illegal), but they do so with the intention of finding vulnerabilities and reporting them to the organization rather than exploiting them for personal gain.

For example, a gray hat hacker might discover a vulnerability in a company's website, exploit it to demonstrate its existence, and then contact the company to inform them of the problem, all without being asked to do so. While their intentions might be good, they're still breaking the law by accessing systems without authorization. Gray hat hacking is controversial. Some people view these hackers as helpful security researchers who aim to make the Internet safer. Others see them as criminals who are breaking the law, regardless of their intentions. Many gray hat hackers have faced legal consequences for their actions, even when they were trying to help.

> **! COMMON MISTAKE: "Good Intentions" Don't Make It Legal**
> Some aspiring security researchers think that hacking into a system to expose and report vulnerabilities is acceptable if their intentions are good. This is wrong. Accessing systems without permission is illegal, regardless of your intentions. Always get written authorization before testing any system you don't own.

For the purposes of this book, we'll focus primarily on the clear distinction between black hat and white hat hackers. If you're interested in hacking as a career, the path forward is clear: become an ethical white hat hacker who operates within legal and ethical boundaries.

Black Hat Hacker Motivations

Now that you understand the different types of hackers, we can explore why black hat hackers do what they do. Not all hackers are the same, and their motivations vary widely. Understanding these motivations helps you understand the threats you face and how to defend against them.

Script Kiddies

At the least sophisticated end of the spectrum, we have what skilled hackers condescendingly call "script kiddies." These are unskilled hackers who use scripts, programs, and tools created by others to carry out attacks. They typically don't understand how these tools work or what they're actually doing; they're just running programs they downloaded. Script kiddies often target websites for defacement, changing the appearance of a website to display their own message or images. Their motivation is often simply to cause mischief, gain bragging rights among their peers, or prove they can do something devious. They're the digital equivalent of vandals spray-painting graffiti.

While script kiddies are the least skilled type of hacker, they can still cause real damage. The tools they use might be simple, but they can still take down websites, deface pages, or cause disruption. And because these tools are readily available on the dark web and elsewhere, the barrier to entry is low. Anyone with basic computer skills and malicious intent can become a script kiddie.

EXAMPLE: Script Kiddie Attack: The TalkTalk Hack (2015)

In October 2015, British telecommunications company TalkTalk suffered a data breach that exposed the personal information of nearly 157,000 customers, including names, addresses, dates of birth, phone numbers, and in some cases, bank account details. The attack was carried out by a group of teenagers, including a 15-year-old boy from Northern Ireland, who used basic SQL injection techniques and readily available scripts that required minimal technical knowledge. The attackers were classic script kiddies: they didn't develop sophisticated exploits. Instead, they used existing tools and scripts downloaded from hacking forums. The breach cost TalkTalk an estimated £77 million in lost revenue, customer compensation, and security improvements. The incident demonstrated that even unskilled attackers using basic techniques can cause significant damage when organizations fail to implement fundamental security measures, such as input validation and regular security testing.

Moving up the ladder from mischief to mission, our next group of hackers is motivated by ideology rather than just seeing what they can get away with.

Hacktivists

Hacktivists are motivated by ideology rather than personal gain. The term combines "hacker" and "activist" and refers to people who use hacking to bring attention to political, social, or environmental causes. A good example would be an environmental activist who opposes fossil fuels and hacks into a large oil company's website to take it down or deface it with environmental messages. Or a political activist who targets government websites to protest policies they disagree with. The hacktivist group Anonymous is perhaps the most well-known example of this type of hacking. Hacktivists view themselves as digital protesters, using hacking as a means for social or political change. However, regardless of their motivations, their actions are illegal. These actions, such as taking down a website, stealing data, or defacing pages, are crimes, even when done for ideological reasons.

EXAMPLE: Hacktivist Attack: Anonymous and Operation Payback (2010)

Anonymous, the decentralized hacktivist collective, launched "Operation Payback" in 2010 to protest against organizations that opposed WikiLeaks. After WikiLeaks released classified U.S. diplomatic cables, several companies, including PayPal, Visa, and Mastercard, stopped processing donations to WikiLeaks. In response, Anonymous coordinated distributed denial-of-service (DDoS) attacks against these companies' websites, temporarily taking them offline. The attacks were ideologically motivated. Anonymous believed these companies were censoring free speech and punishing WikiLeaks for exposing government secrets. The operation demonstrated how hacktivists use cyberattacks as a form of digital protest, targeting organizations they perceive as acting against their values. While the attacks caused temporary disruptions and financial losses, the primary goal was to make a political statement rather than to steal data or cause permanent damage. This example illustrates how hacktivists differ from other threat actors: their motivation is ideological rather than financial, and their attacks are often designed to generate publicity for their cause.

While hacktivists target organizations for political reasons, state-sponsored hackers operate on a global scale with the full backing of their governments.

State-Sponsored Hackers

State-sponsored hackers are among the most sophisticated and dangerous threat actors. These are hackers who work for a government, either as part of a governmental agency, within that government's military, or as contractors. They hack with direction, support, and often training from their government. State-sponsored hackers are what we refer to as advanced persistent threats (APTs), as they pose a continual threat that requires ongoing diligence. Governments are constantly conducting cyberoperations against other governments for a variety of reasons, including espionage, sabotage, or as part of broader geopolitical conflicts.

State-sponsored hackers typically have significant resources at their disposal: advanced tools, extensive training, and substantial budgets. They may spend months or even years inside a target's network, quietly gathering intelligence. Their operations are often highly sophisticated and difficult to detect. State-sponsored hacking has become a major component of modern geopolitical conflict. Nations employ cyber operations to spy on one another, disrupt critical infrastructure, steal intellectual property, and gain strategic advantages.

EXAMPLE: State-Sponsored Attack: APT28 and the DNC Hack (2016)

In 2016, the Democratic National Committee (DNC) suffered a major cyber breach attributed to APT28 (also known as Fancy Bear), a hacking group linked to Russia's military intelligence agency GRU. The attackers used spear-phishing emails to gain initial access to the DNC network, then spent months conducting reconnaissance, stealing emails, and exfiltrating sensitive documents. The stolen data was later released through WikiLeaks and other platforms during the 2016 U.S. presidential election, causing significant political disruption.

This attack exemplifies state-sponsored hacking: It was sophisticated, well-resourced, and conducted over an extended period with clear geopolitical objectives. Unlike financially motivated cybercriminals, APT28's goal was to influence political outcomes and gather intelligence. The attackers employed advanced techniques, including custom malware, credential harvesting, and meticulous operational security, to evade detection. State-sponsored hackers, such as APT28, represent advanced persistent threats because they have government backing, substantial resources, and long-term strategic objectives that extend beyond immediate financial gain.

Cyberterrorists

Cyberterrorists are hackers whose goal is to cause harm to infrastructure or civilians through cyberattacks. They use hacking as a tool of terrorism, seeking to create fear, cause disruption, or inflict large-scale damage. The distinction between cyberterrorists and other hackers lies in their intent to cause widespread harm and instill terror. They might target critical infrastructure, such as power grids, water treatment facilities, or transportation systems. They may seek to disrupt emergency services or healthcare systems. Their goal isn't just to steal data or make money, but to cause real-world harm and instill fear. While large-scale cyberterrorism hasn't been as prevalent as some analysts feared 20 or 30 years ago, the threat remains real. As our critical infrastructure becomes increasingly connected and dependent on digital systems, the potential for cyberterrorism grows.

EXAMPLE: Cyberterrorism Threat: Potential Attacks on Critical Infrastructure

While large-scale cyberterrorism attacks causing physical harm remain rare, the threat is real. Cyberterrorists could target power grids, water treatment facilities, transportation systems, and healthcare networks. The 2021 Colonial Pipeline ransomware attack disrupted fuel supplies across the eastern U.S., demonstrating critical infrastructure vulnerability. Though financially motivated, it showed how attackers could manipulate industrial control systems (ICS) or SCADA systems to cause equipment failures, chemical releases, or power outages. The 2015 Ukraine power grid attack, which left 230,000 without electricity, proved such attacks are feasible. As infrastructure becomes increasingly connected, protecting critical systems is a top national security priority.

Finally, some cybercrime isn't driven by ideology or espionage, but by pure business.

Hacking-as-a-Service (HaaS)

Finally, we have a relatively new phenomenon: hacking-as-a-service. This is where hacking has literally become a business model. Skilled black hat hackers sell their services to individuals or organizations that want to conduct cyberattacks but lack the technical skills to do so. These are hacking groups, often operating from countries with limited cybersecurity law enforcement, that offer a range of services, including DDoS attacks, data theft, ransomware deployment, credential theft, and more. Customers can simply pay for the service they want, and the HaaS provider carries out the attack.

This model has turned cybercrime into a capitalist endeavor, making sophisticated attacks accessible to anyone with money and malicious intent, regardless of technical skill. HaaS has significantly lowered the barrier to entry for cybercriminals and contributed to the dramatic increase in the volume and variety of cyberattacks we've seen in recent years. Ransomware-as-a-Service (RaaS), a specific type of HaaS, has become particularly prevalent, which we'll explore in more detail in Chapter 5 when we discuss malware threats.

The Hacking Methodology

With a clear understanding of who hackers are and what drives them, let's explore how they actually carry out their attacks. Both black hat and white hat hackers follow a similar method, a systematic approach to gaining access to systems. Understanding this methodology helps you defend against attacks.

Since this isn't an ethical hacking guide, we'll keep it brief. We won't get into the technical details of specific tools or techniques. Instead, we'll walk through the framework that hackers use, giving you a 10,000-foot view of the hacking process.

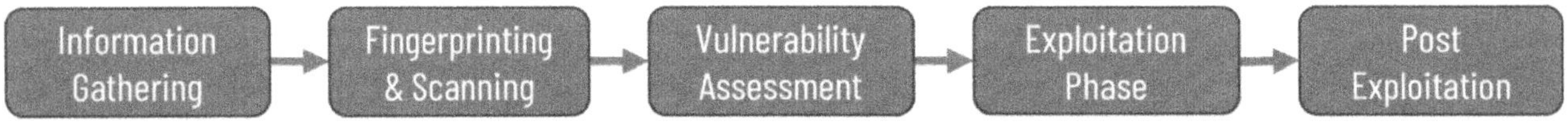

Figure 2.1: The Hacking Methodology

Step 1: Information Gathering

Every hacking operation begins with information gathering. Hackers refer to this as open-source intelligence gathering, or OSINT, which involves the collection and analysis of publicly available information from sources such as websites, social media, public records, and press releases. Hackers, whether black hat or white hat, start by examining publicly accessible information about their target, gathering intelligence without ever touching the target's systems.

They examine websites, looking for information about the organization's structure, employees, technologies used, and contact information. They search social media profiles to learn about employees and their roles. They review job postings to understand which technologies the organization uses. They review press releases, news articles, and public filings. They search for any information that might be useful in planning an attack.

This phase might also include social engineering. For example, if a website lists employees' contact information, a hacker might call or email them, impersonating someone else, to gather additional information. They might pretend to be a vendor, a customer, or even another employee to extract useful details. The information gathering phase is crucial because it helps hackers understand their target and identify potential entry points. The more information they can gather, the more effective their attack is likely to be.

Step 2: Fingerprinting & Scanning

Once hackers have gathered publicly available information and identified potential entry points, they move on to fingerprinting and scanning the network. This is where they begin to interact directly with the target's systems, albeit still in a relatively passive way.

The purpose of this phase is to identify the IP address schema and the operating systems running on the network. They want to understand the network's layout and architecture. What devices are connected? What operating systems are they running? What services are available? What ports are open? Hackers use various scanning tools to probe the network and gather this information. They're essentially creating a map of the target's network, identifying all the devices and understanding how they're configured. This gives them a clear picture of the attack surface they're working with.

Step 3: Vulnerability Assessment

With a map of the network in hand, hackers move on to the vulnerability assessment phase. This is where they scan devices, switches, routers, servers, workstations, and other systems, looking for known vulnerabilities.

They're looking for unpatched software, misconfigured systems, weak passwords, unnecessary services running, and any other security weaknesses. They use automated vulnerability scanning tools that check systems against databases of known vulnerabilities. The goal is to identify which systems have weaknesses that could be exploited. Not every vulnerability will be exploitable in practice, but this phase helps hackers identify the most promising targets and attack vectors.

Step 4: Exploitation

This is where things get interesting, and where black hat and white hat hackers really demonstrate their skills. The exploitation phase is where hackers attempt to actually gain access to systems by exploiting the vulnerabilities they've identified.

They look for specific exploits that match the vulnerabilities they've found. An exploit is essentially a tool or technique that takes advantage of a specific vulnerability. For example, if they've identified that a server is running an outdated version of software with a known vulnerability, they'll look for an exploit that targets that specific vulnerability.

This phase requires skill, creativity, and often a lot of trial and error. Not every exploit works in every situation. Hackers need to understand the systems they're targeting and adapt their approach based on what they encounter. For white hat hackers, this is often the most enjoyable part of the job; it's like solving a complex puzzle. For black hat hackers, this is where they gain the access they need to carry out their malicious objectives.

Step 5: Post-Exploitation

Once a hacker has successfully exploited a vulnerability and gained access to a system, they enter the post-exploitation phase. This is where the real damage occurs, and where the paths of black hat and white hat hackers diverge significantly.

Black Hat Post-Exploitation

Once a black hat hacker has successfully exploited a vulnerability and gained access to a system, they enter the post-exploitation phase, where their true malicious intent becomes evident. This is where the real damage occurs. Unlike the earlier reconnaissance and exploitation phases, which are largely technical exercises, post-exploitation is where hackers achieve their ultimate objectives: whether that's financial gain, espionage, sabotage, or simply causing chaos. The specific actions they take during this phase depend on their motivations, skills, and targets, but they generally follow a pattern of consolidating their access, expanding their reach, and extracting value from the

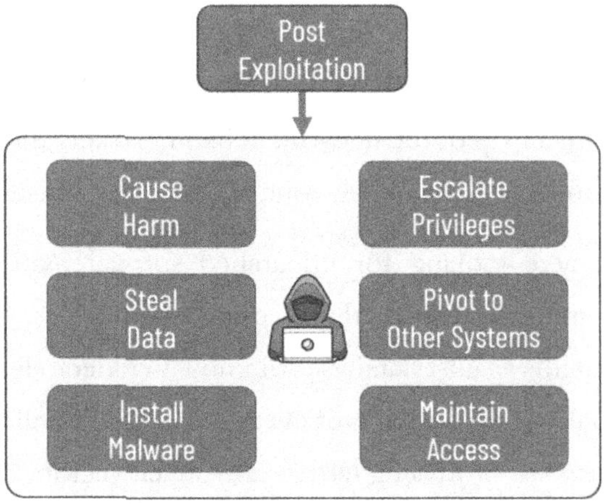

Figure 2.2: Black Hat Post-Exploitation

compromised system. A black hat hacker who's gained access to a system will typically pursue several objectives:

- **Escalate Privileges**: If they've gained limited access, they'll attempt to exploit vulnerabilities to elevate their access rights from a standard user to an administrator or root account, thereby gaining complete control over the system.
- **Maintain Access**: They'll install backdoors or create new user accounts to ensure they can return to the system even if their initial entry point is discovered and closed.
- **Pivot to Other Systems**: They'll use their access to one system as a launching point to attack other systems on the network, gradually expanding their access.
- **Steal Data**: They'll search for and exfiltrate valuable data, including customer information, financial records, intellectual property, credentials, or whatever their target is.
- **Install Malware**: They might install ransomware, spyware, or other malicious software to further their objectives.
- **Cause Harm**: Depending on their motivation, they may delete data, disrupt operations, or cause other forms of damage.

The black hat hacker's goal is malicious: to profit from the attack, cause damage, steal information, or achieve whatever objective motivated them to hack the system in the first place.

White Hat Post-Exploitation

A white hat hacker who's successfully exploited a vulnerability takes a very different approach. They document what they've done, how they did it, and what access they gained. They might explore the extent of access they can achieve:

- Can they escalate privileges?
- Can they pivot to other systems?
- Can they access sensitive data?

However, they do all of this within the boundaries defined in their pre-engagement agreement. They don't actually steal data, cause damage, or do anything malicious. They're simply proving what a malicious hacker could do if they exploited the same vulnerability. The ethical hacker will then follow all the steps in the hacking methodology until post-exploitation, just as a black hat hacker would. Once they reach post-exploitation, they'll write a report of their findings, including the vulnerabilities they discovered, those they successfully exploited, and recommendations for remediating them. This is where the goals diverge: malicious hackers aim to cause harm, while ethical hackers aim to provide recommendations to enhance security.

The Ethical Hacking Framework: Adding Professional Guardrails

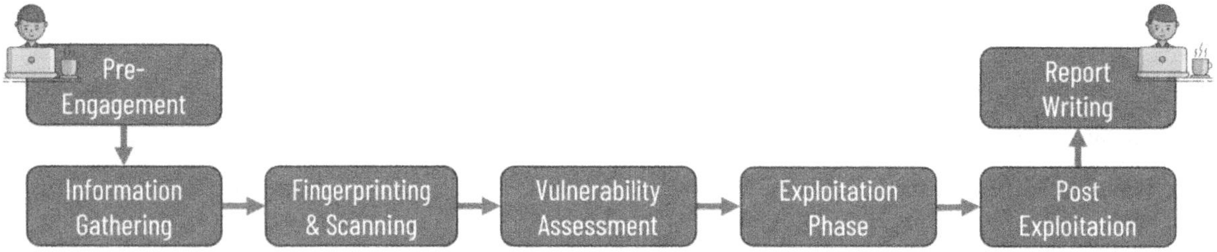

Figure 2.3: The Ethical Hacking Framework

While ethical hackers use the same core five-step methodology, their work is bookended by two crucial, non-technical steps that separate them from criminals: Pre-Engagement and Reporting.

Step 0: Pre-Engagement

Before white hat hackers begin any hacking activities, they follow a critical step that black hat hackers skip entirely: the pre-engagement process. This is where the ethical hacker and client work together to define the scope, rules of engagement, objectives, and legal boundaries of the security assessment before any testing begins.

Before a white hat hacker starts an ethical hacking engagement, they'll sit with the stakeholders and have a pre-engagement meeting to determine and receive the following:

- The Scope of the Ethical Hacking Engagement
- Black, Gray, or White Box Test
- Ethical Hacking Consent Legal Agreement

The Scope of the Ethical Hacking Engagement

First and foremost, the stakeholders must provide the ethical hackers with the overall scope of the engagement. Whether they want a vulnerability assessment or a full-blown penetration test, and what systems and networks they want assessed. Additionally, the systems and networks to be excluded from the assessment are selected; it's common for companies to exclude mission-critical systems because penetration testing can break them. The stakeholders then determine whether the company wants the ethical hackers to perform social engineering attacks, and, lastly, they set the time frame and number of days for the engagement.

Black, Gray, or White Box Test

The stakeholders will then determine what information the company will share to determine if the ethical hacking will be a black, gray, or white box test:

- **Black Box**: With a black box test, the ethical hacker isn't given any internal knowledge of the network or systems; the goal is to hack the network and systems from an outsider's perspective.
- **Gray Box**: With gray box testing, ethical hackers are given limited internal access and information to simulate a more focused attack on a network or system. The goal is to simulate a hacker who's already breached the perimeter network or an insider threat with limited access and knowledge.
- **White Box**: With white box testing, ethical hackers are given all information regarding the systems and networks. They may also be given high-level privileges to the network as well. This allows the ethical hackers to perform a comprehensive security assessment for the company.

Ethical Hacking: Consent & Legal Agreement

Lastly, before the work starts, an ethical hacking consent is drafted based on the scope of the engagement and is agreed upon and signed by both parties. This provides the ethical hacking team with the legal and formal approval to perform the requested vulnerability assessment or penetration test.

Step 6: Report Writing (White Hat Only)

The final phase for white hat hackers is report writing. After completing their assessment, ethical hackers compile a comprehensive report detailing all their findings. This report typically includes:

- An executive summary for non-technical stakeholders
- A detailed list of all vulnerabilities discovered
- Risk ratings for each vulnerability
- Evidence of vulnerabilities (screenshots, logs, etc.)
- For penetration tests, documentation of which vulnerabilities were successfully exploited
- Detailed recommendations for remediation
- Prioritization of fixes based on risk

The goal of this report is to help the organization improve its security. It provides a roadmap for fixing vulnerabilities and strengthening defenses. This is the fundamental difference between white hat and black hat hackers: the white hat hacker's goal is to improve IT security, while the black hat hacker's goal is to exploit it for malicious purposes.

Bringing It All Together

Understanding hackers is essential to defending against them. This chapter has taken you beyond Hollywood's dramatized portrayals to reveal the reality: hacking is a methodical, time-consuming process that requires patience and systematic thinking.

The distinction between black hat and white hat hackers isn't about technical skill—both often use the same tools and techniques. The difference lies in permission, intent, and ethics. Black hat hackers exploit vulnerabilities for malicious purposes, while white hat hackers identify and fix those same vulnerabilities to strengthen defenses.

Black hat hacker motivations are diverse and evolving. From script kiddies causing mischief to state-sponsored actors conducting sophisticated espionage operations, each threat actor brings different capabilities and objectives. The rise of hacking-as-a-service has democratized cybercrime, allowing even non-technical criminals to launch sophisticated attacks. Understanding these motivations helps you anticipate threats and tailor your defenses accordingly.

The hacking methodology (information gathering, scanning, vulnerability assessment, exploitation, and post-exploitation) provides a roadmap for both attack and defense. When you understand the systematic approach hackers use, you can implement security measures at each stage to disrupt their progress. Every phase of the methodology represents an opportunity to detect and stop an attack before it succeeds. As you continue through this book, you'll learn how to defend against the techniques described in this chapter. The next chapter on networking will show you the primary attack surface that hackers target and how to protect it.

Key Takeaways

- **Real hacking is methodical and time-consuming**, taking hours or days rather than the dramatic minutes portrayed in movies. Ethical hacking certifications often allow anywhere from 24 hours up to several days to complete their tests.
- **Black hat hackers are criminals** who exploit vulnerabilities for malicious purposes, while white hat hackers are ethical professionals who use the same techniques with permission to identify and fix security weaknesses.
- **Black hat hacker motivations vary widely**, from script kiddies seeking mischief to state-sponsored actors conducting espionage, hacktivists pursuing ideological goals, cyberterrorists seeking to cause harm, and hacking-as-a-service providers offering criminal services for hire.

- **The hacking methodology follows a systematic process**: information gathering (OSINT), fingerprinting and scanning, vulnerability assessment, exploitation, and post-exploitation. Black hat hackers use these methods to cause harm, whereas white hat hackers apply them to help stakeholders defend against and disrupt attacks at each stage.
- **White hat hackers meet with stakeholders** initially to define the scope of the operation and obtain legal authorization before beginning any testing, and then to conclude the operation by submitting detailed reports that provide organizations with actionable recommendations to improve their security posture.

Quick Action: Reduce Your Digital Footprint

In this chapter, you learned how hackers use OSINT (open-source intelligence gathering) to research their targets before attacking. Everything you share publicly gives attackers valuable information they can use to craft convincing phishing attacks or social engineering schemes. The best defense against reconnaissance is making yourself a harder target to research. So, in this quick action, tighten what you share online by reviewing your social media privacy settings, removing or hiding sensitive personal details, and searching for yourself online so you can remove or opt out of any unwanted public listings.

Protect Your Privacy

Review Your Social Media Privacy Settings

- Check what information is publicly visible on LinkedIn, Facebook, Twitter, and other platforms.
- Consider making your profiles private or limiting what strangers can see.
- Remove or hide personal information such as your home address, phone number, birth date, and family member names.

Search for Yourself Online

- Google your name, email address, and phone number.
- See what information about you is publicly available.
- Consider using opt-out services for data broker websites that list your personal information.
- Be mindful of what you share publicly going forward.

Remember: Every piece of information an attacker can gather during reconnaissance makes their job easier. Limiting your public information exposure is a simple but effective defense.

Chapter 3. Networking 101

Introduction

When we think about hackers, the number one attack vector they use to break into IT systems is computer networks. As this book progresses, we'll discuss common network-based attacks and essential security measures. But to understand them, you first need to know how networks operate. Think about it this way: If you want to protect your home from burglars, you need to understand how they might break in, through windows, doors, or other entry points. The same logic applies to cybersecurity. Without understanding how networks function, the security concepts we'll discuss later simply won't make sense.

This chapter serves as your crash course in computer networking. We'll cover the fundamental concepts you need: basic networking rules and protocols, local area networks (LANs) versus wide area networks (WANs), Media Access Control (MAC) and Internet Protocol (IP) addresses, switches and routers, the Internet Protocol suite, commonly known as TCP/IP, and how the Internet actually works. These aren't just abstract technical concepts; they form the foundation for understanding the network security topics that follow.

What Is a Computer Network?

Figure 3.1: Computer Network

In its simplest form, a network is two connected computers sharing resources with one another. While the diagram above depicts two desktop computers, in 2025 and beyond, when we think about computers, we can include anything ranging from smartphones to tablets to IoT devices within our homes.

When I say "sharing resources," I'm talking about things like files, printers, Internet connections, or even processing power. If you've ever sent a document from your laptop to a wireless printer, streamed a movie from your computer to your smart TV, or accessed files stored on one device from another, you've used a network.

> **EXAMPLE**: When you print a document wirelessly from your laptop to your home printer, your laptop sends the document data across your home network to the printer. The printer receives the data and prints it, all without any physical connection between the devices.

While our simple definition involves just two computers, real-world networks are typically much more complex. A home network might have a dozen devices connected, while a business network might have hundreds or even thousands. Regardless of the scale, the fundamental concept remains the same: a network is about connecting devices so they can share resources and communicate.

Two Main Aspects

In regards to any network, there are two main aspects: the physical and logical connections. The physical connection is the wires and cables that connect a network together, as well as the wireless signals propagating throughout the air between wirelessly connected devices. The logical connection, on the other hand, refers to the data being transported back and forth across the physical connection, whether via cables or Wi-Fi.

Think of it like a highway system: the physical connection is the roads themselves, while the logical connection is the cars traveling on those roads. You need the roads for the cars to travel, but the roads themselves aren't doing anything useful without the cars. Practically speaking, networks always have both a physical layer (cables and wireless signals) and a logical layer (the data traveling across them), and attackers can target either, so this distinction matters for security.

Basic Networking Rules

With the physical and logical aspects defined, we turn to the basic networking rules that allow networks to operate. Every network in the world has procedures it must follow to function correctly and to send and receive data effectively. In the world of IT networking, we call these rules communication protocols.

Without these rules, networks would be complete chaos. Imagine if there were no rules governing how devices communicate, one device might try to send data in one format while another device expects a completely different format. It would be like trying to have a conversation where one person speaks English, another speaks Mandarin, and a third speaks Arabic, with no translator present. Nothing would get accomplished.

Communication protocols solve this problem by establishing a common language and a set of procedures that all devices on a network must follow. Specifically, they're the rules that govern how devices communicate by ensuring data is formatted, transmitted, and received correctly. These protocols dictate everything from how data is formatted and packaged to how devices identify themselves on the network, and even how errors are detected and corrected.

Origin & Destination Addresses

Additionally, computers within a network need to know the origin and destination addresses when communicating. In the networking world, we call these IP addresses and MAC addresses, which you'll learn about later in this chapter.

Think about this concept in terms of sending a physical letter through the postal service. For your letter to reach its intended recipient, you need to write two addresses on the envelope: the destination address (where you want the letter to go) and the return address (where the letter came from). Computer networks work the same way; every piece of data carries a destination address (where it's going) and a source address (where it came from), so it can be delivered and a response can be returned.

Beyond routing, the origin address also serves another critical purpose: accountability. If a device on the network is behaving maliciously, the origin address allows network administrators to identify which device is causing the problem.

Computer Protocols

Having established the importance of addresses, let's dive deeper into computer networking protocols. Computer networking protocols allow you to communicate with other devices on a network. If we take a step back and talk about protocols in general, protocols are nothing more than rules.

To illustrate this, let's use a simple driving example. Let's say that cars driving on the street are packets of data going across a network, and there's an intersection. When people driving cars obey the rules of the road, and there aren't any accidents, data flows smoothly throughout the network with no issues. However, what happens if a car runs a red light, causing a major accident? That's going to cause a traffic jam. Looking at it from the perspective of a computer network, data won't be able to travel across the network until the accident is resolved and traffic is flowing again. In networking terms, protocols are those rules of the road: when they're followed, data flows efficiently; when they're broken, either accidentally or maliciously, communication breaks down.

From a cybersecurity perspective, understanding protocols is crucial because many attacks exploit weaknesses in these protocols or involve violating protocol rules in specific ways to achieve malicious goals.

Example TCP/IP Protocols

With that foundation in place, here are some real-world examples of protocols. TCP/IP is the most widely used suite of network protocols in the world. In fact, it's the protocol suite on which modern networks operate. We'll discuss it later in the chapter, but as a primer, let's look at a few of the more prominent protocols in the TCP/IP suite.

Web Communication: HTTP & HTTPS

Just about everybody browses the web daily. When we do so, our web browser uses the HTTP or HTTPS protocol to communicate with web servers to retrieve the web pages we want to view.

HTTP stands for Hypertext Transfer Protocol, and it serves as the foundation for data communication on the World Wide Web. Every time you type a web address into your browser and hit enter, your browser sends an HTTP request to a web server asking for the web page you want to view. The server then responds with the web page data, which your browser displays on your screen.

HTTPS is the safe version of HTTP—the "S" stands for "Secure." It works the same way as HTTP, but it encrypts the data being transmitted between your browser and the web server. This encryption protects sensitive information, such as passwords, credit card numbers, and personal data, from being intercepted by hackers. You can tell if a website is using HTTPS by looking for the padlock icon in your browser's address bar.

The difference between HTTP and HTTPS is critical from a security standpoint. When you're using HTTP, all the data you send and receive is transmitted in plain text, meaning anyone who intercepts it can read it. HTTPS, on the other hand, is like sending a letter in a sealed, tamper-proof envelope. Today, most sites default to HTTPS; still, make a habit of checking for the padlock symbol on your web browser before entering passwords, payment details, or other sensitive data.

Email: POP3, SMTP, & IMAP

Email is the preferred method of communication in many businesses. Email clients, such as Outlook and Thunderbird, and email servers use POP3, SMTP, and IMAP protocols to send and receive emails.

SMTP, which stands for Simple Mail Transfer Protocol, is used for sending emails. When you compose an email and hit send, your email client uses SMTP to transmit that email to your email server, which then uses SMTP to forward it to the recipient's email server.

POP3, which stands for Post Office Protocol version 3, is one method for receiving emails. When you use POP3, your email client downloads emails from the server to your device and typically deletes them from the server. It's like going to a post office box, collecting all your mail, and taking it home with you.

IMAP, which stands for Internet Message Access Protocol, is another method for receiving emails, but it works differently from POP3. With IMAP, your emails stay on the server, and your email client just displays them. This allows you to access your emails from multiple devices—your phone, laptop, and tablet can all show the same emails because they're all connecting to the same server.

EXAMPLE: If you check your email on both your phone and laptop and see the same messages on both devices, you're using IMAP. If emails disappear from your phone after you read them on your laptop, you're using POP3.

Name Resolution: DNS

When you want to visit a website, you simply type in the domain name of that website, such as YouTube.com, into your web browser. You don't need to know the website's IP address, just the domain name. However, your computer does need to know it. To send an HTTPS request to YouTube, your browser needs to know YouTube's IP address, for example, 208.65.153.238. That's where the Domain Name System (DNS) comes in.

Think of DNS as the phone book of the Internet; it resolves human-friendly domain names into the computer-friendly IP addresses needed to make a connection, so you don't have to remember them. Just as you can look up a person's name in a phone book to find their phone number, DNS allows your computer to look up a domain name to find its corresponding IP address. When you type a domain name into your browser, your computer sends a DNS query to a DNS server asking, "What's the IP address for this domain name?" The DNS server looks it up in its database and responds with the IP address. Your computer then uses that IP address to connect to the website. This entire process happens in milliseconds.

KEY POINT: DNS is the phone book of the Internet—it translates human-friendly domain names (like google.com) into IP addresses (like 172.217.164.46) so your computer knows where to connect.

From a security perspective, DNS is a critical component of the Internet infrastructure, making it a prime target for attackers. Hackers can manipulate DNS to redirect users to malicious websites, intercept communications, or completely take down websites. We'll discuss DNS-based attacks, including DNS spoofing, in Chapter 6.

These are just a few examples of commonly used TCP/IP protocols. There are dozens of other protocols in the TCP/IP suite, each serving a specific purpose. The protocols we've discussed here, HTTP/HTTPS, email protocols, and DNS, are simply the ones you're most likely to encounter in your daily Internet usage.

LANs & WANs

With a basic understanding of protocols, let's discuss the differences between LANs (local area networks) and WANs (wide area networks)

Local Area Networks (LANs)

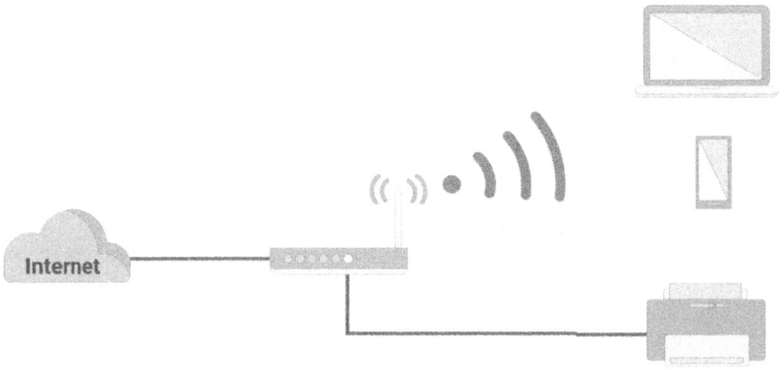

Figure 3.2: Local Area Network (LAN)

A LAN is a small network confined to a limited geographical region, such as a single room, a single building, or a small group of buildings. The hallmark feature of a LAN is that all the devices are directly connected. Some examples of a LAN would be a home network, a small business's office network, or an academic department's network within a university.

When devices are directly connected, they're part of the same network segment and can communicate with each other without needing to go through external networks or the Internet. For example, in your home network, all your devices, your laptop, smartphone, smart TV, and gaming console, are all directly connected to your home router. They can all communicate directly with each other without their data leaving your home network.

Wide Area Networks (WANs)

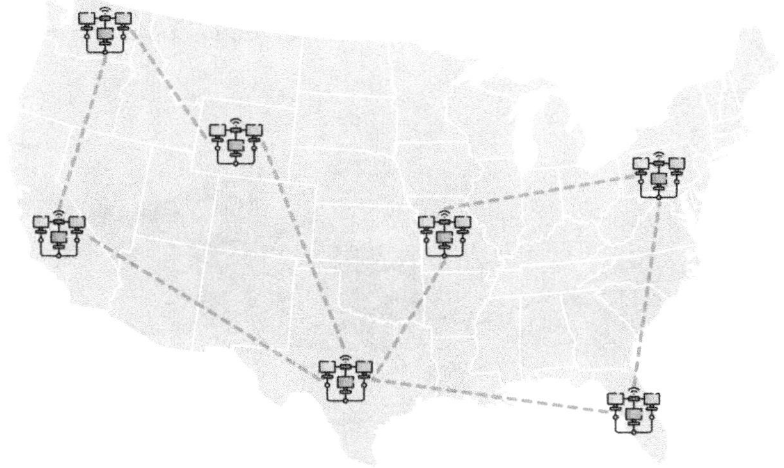

Figure 3.3: Wide Area Network (WAN)

On the flip side, we have WANs, which are much larger than LANs. WANs typically span a large geographical region, connecting distant LANs together. For example, a university campus's WAN connects each academic department's LANs into a larger network. Another example would be a large corporation with offices across the continental United States. While each office has its own separate LAN, the company can create a corporate WAN to connect the offices, which would allow all offices to communicate securely with one another. If you're wondering how big a WAN can be, the largest one in the world is the Internet itself.

> **EXAMPLE:** A national bank with branches in every state uses a WAN to connect all its branch LANs together. When you deposit a check at a branch in California, the transaction data travels across the bank's WAN to update your account, which might be managed by servers in New York.

The Internet is the ultimate example of a WAN. It's a massive network of networks, connecting millions of LANs around the world. When you access a website hosted on a server in another country, your data travels across this global WAN, passing through numerous networks and routers to reach its destination. In short, LANs connect devices locally (like your home network), while WANs connect networks across distances (like the Internet). This distinction clarifies where data travels and where risks tend to appear.

MAC & IP Addresses

We'll now discuss MAC and IP addresses, building on our earlier section on origin and destination addresses.

MAC Addresses

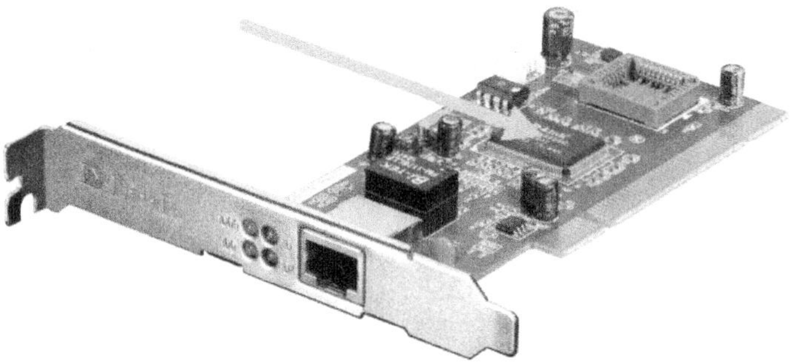

Figure 3.4: Network Interface Card (NIC)

Every single computer in the world that's connected to a network has a network interface card (NIC). The NIC is what allows you to connect a device to a network. It could be a wired NIC or a wireless one that's in a tablet, a laptop, a smartphone, or an IoT device. All NICs have two commonalities:

1. They provide a means to connect a device to a network.
2. Each has a unique MAC address.

A MAC (Media Access Control) address is a 48-bit identifier that looks something like this: 00:1A:2B:3C:4D:5E. It's typically displayed as six pairs of hexadecimal digits separated by colons or hyphens.

The MAC address is physically burned into the card's ROM chip; that's why it's commonly referred to as a computer's physical address. MAC addresses are used for communication between devices on the same LAN via switches, which we'll discuss in detail later in the chapter. When your laptop wants to send data to your printer on your home network, it uses the printer's MAC address to ensure the data reaches the right device.

Because MAC addresses are burned into the hardware, they're designed to be permanent. However, in reality, it's possible to change a MAC address through software via a technique called MAC spoofing, which is something hackers can do to hide their identity or impersonate other devices.

> **KEY POINT**: MAC addresses are physical addresses burned into network hardware. They're used for local communication within a LAN via switches, but they can be spoofed by attackers.

IP Addresses

Now let's discuss IP (Internet Protocol) addresses. IP addresses aren't burned onto the ROM chip on the NIC, so they're not physical addresses. Instead, they're logical addresses, meaning they're logically assigned within the operating system by the person who owns the device. That could be you or me within our home network, an IT administrator within a business, or via a DHCP server that automatically assigns IP addresses when devices connect to a network.

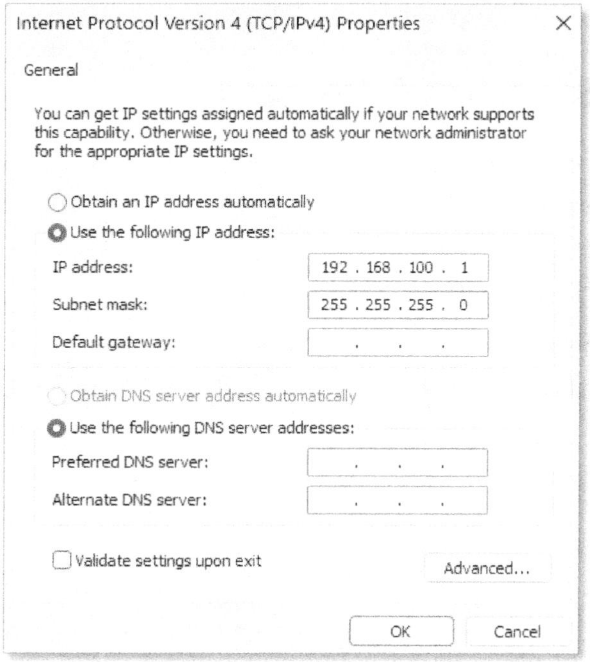

Figure 3.5: Windows TCP/IP Settings

Here's the hallmark difference between MAC and IP addresses: MAC addresses are designed for local communication within a LAN using switches, while IP addresses are designed for global communication across networks using routers. MAC addresses help devices find each other on the same network segment; they're like apartment numbers within a building. IP addresses are hierarchical and routable, meaning they contain information about which network a device belongs to; they're like full mailing addresses that include the street address, city, state, and zip code.

EXAMPLE: Think of MAC addresses as apartment numbers (useful within a building) and IP addresses as full mailing addresses (needed to reach different buildings). You need both: the full address gets mail to the right building, and the apartment number gets it to the right unit.

When you send data across the Internet, both types of addresses are used. The IP address gets your data from your network to the destination network, and then the MAC address gets it to the specific device on that destination network.

There are two versions of IP addresses in use today: IPv4 and IPv6. IPv4 addresses look like this: 192.168.1.1, four numbers separated by periods, with each number ranging from 0 to 255. IPv6 addresses are much longer and were developed because we were running out of available IPv4 addresses as more and more devices connected to the Internet. Put simply, MAC addresses handle local delivery within a LAN (via switches), while

IP addresses enable routing across networks (via routers). Both are essential because they serve different purposes.

Switches & Routers

Now that you understand MAC and IP addresses, let's explore the two types of devices that use them: switches and routers.

Switches

Switches connect devices within a LAN using MAC addresses. They operate by maintaining a MAC address table that maps each connected device's MAC address to a specific physical port on the switch.

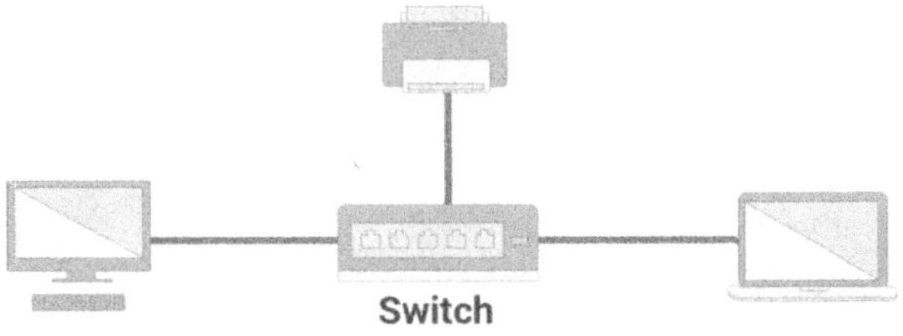

Figure 3.6: Network Switch

When data needs to be sent to a device, the switch looks up the destination MAC address in its table, identifies the physical port to which that device is connected, and forwards the data only to that specific port. This intelligent forwarding reduces unnecessary network traffic and improves performance.

> **EXAMPLE**: When you print a document from your laptop on your office network, the switch looks up your printer's MAC address, identifies which port it's connected to, and sends the print data only to that port, not to your coworkers' computers or other devices.

Routers

While switches manage traffic within a LAN, routers connect different networks together. They route traffic between networks using IP addresses and routing protocols. Routing protocols use metrics and algorithms to determine the best path to get a packet of information from one network to another, whether across a small WAN or across the Internet via thousands of routers.

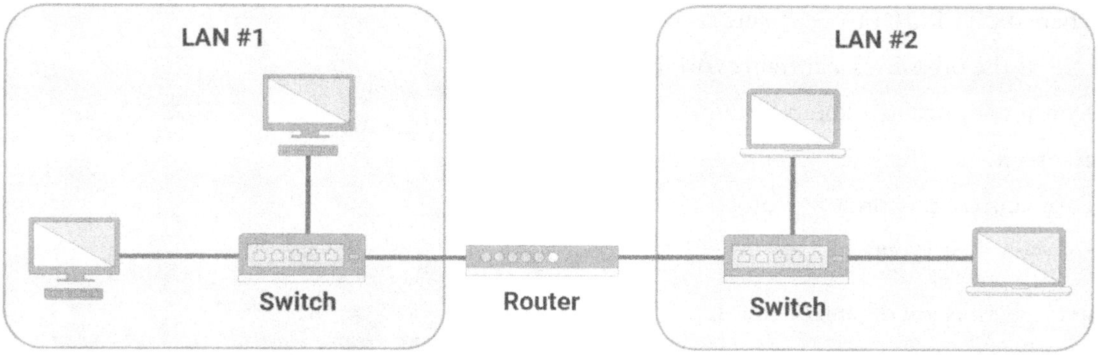

Figure 3.7: Router Connecting Two Networks

Additionally, routers maintain routing tables that contain information about how to reach different networks. When a router receives a packet of data, it examines the destination IP address, consults its routing table, and forwards the packet toward its destination. Each router along the path makes an independent decision about the best next hop.

> **EXAMPLE**: When you visit a website, your request leaves your home network through your router, which forwards it to your ISP's router, which then forwards it to other routers across the Internet until it reaches the web server's network. The response travels back through a similar path to reach your device.

Furthermore, from a security perspective, routers often serve as the first line of defense for a network. They typically include firewall functionality that filters incoming and outgoing traffic, blocking potentially malicious connections before they reach devices on your network.

> **KEY POINT**: Switches use MAC addresses to forward data within a LAN. Routers use IP addresses to forward data between different networks.

> **COMMON MISTAKE**: Don't confuse the roles. Switches handle local traffic using MAC addresses, while routers handle traffic between networks using IP addresses.

The TCP/IP Protocol Suite

Now that you understand the physical infrastructure of networks (including switches, routers, and how data travels through them), let's return to the protocols that govern communication across this infrastructure. We've explored several individual protocols, including HTTP/HTTPS for web browsing, SMTP/POP3/IMAP for email, and DNS for name resolution. But how do all these protocols work together?

That's where the TCP/IP protocol suite comes in. Think of TCP/IP as the organizational framework that houses all these protocols. Just as a library organizes books into different sections (fiction, nonfiction, reference), TCP/IP organizes networking protocols into distinct layers based on their functions.

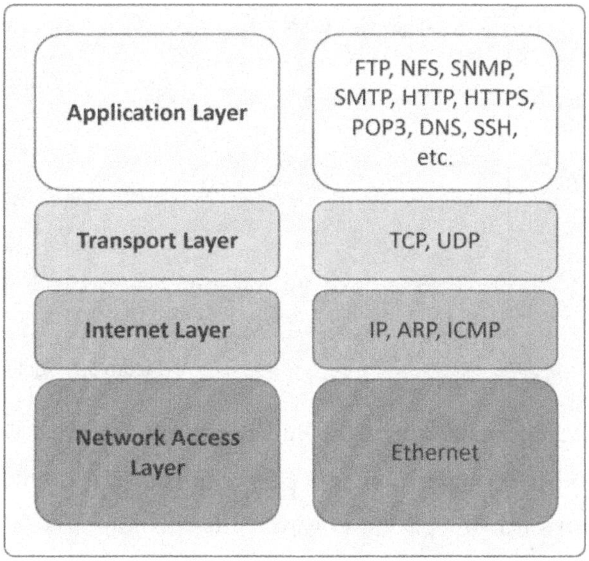

Figure 3.8: TCP/IP Protocol Suite

Computer networks are organized into layers, much like sending a letter involves multiple steps. You write the message (content), put it in an envelope (packaging), write addresses on it (routing information), and hand it to the postal service (physical delivery). Each layer handles a specific job.

The TCP/IP protocol suite works the same way. The protocols you've already learned about, HTTP, SMTP, and DNS, all reside in specific layers of this suite. Understanding how these layers work together helps you see the complete picture of network communication.

The Four Layers of TCP/IP

Regarding its design, TCP/IP is broken up into four layers. Each layer serves a specific purpose in the networking communication process, where specific protocols reside. Below are highlights of each layer's responsibilities:

- **Application Layer**: provides an interface to access network resources.
- **Transport Layer**: ensures data is delivered error-free and in sequence.
- **Internet Layer**: provides IP addressing and routing services.
- **Network Access Layer**: ensures data is delivered to the proper device on a LAN using MAC addresses.

Protocol Selection & Purpose

Networks use different protocols for different purposes. Some protocols, like those used for web browsing and email, prioritize accuracy. Every piece of data must arrive correctly and in order, even if it takes a bit longer. Other protocols, such as those used for video streaming or online gaming, prioritize speed—it's okay if a few pieces of data are lost, as long as the stream continues to flow smoothly. Think of it like sending an important email versus watching a live video stream. For the email, every single word must be perfect (accuracy). For the live video, a few dropped pixels you barely notice are an acceptable trade-off for a smooth, uninterrupted stream (speed).

So when we're talking about computer networks, whether they be LANs or WANs, just about every network uses TCP/IP. And from a cybersecurity perspective, that's important because the protocols in the TCP/IP suite can serve as attack vectors that hackers can exploit to compromise devices and systems. In Chapter 6, we'll discuss protocol spoofing attacks, specifically ARP poisoning and DNS spoofing.

Additionally, the layered architecture of TCP/IP has security implications. Attacks can target any layer of the stack. Understanding these layers helps you understand where vulnerabilities exist and how different types of attacks work. In practice, TCP/IP is a four-layer suite that enables Internet communication, with each layer carrying distinct responsibilities and its own potential weaknesses.

> ! **COMMON MISTAKE**: Don't assume all network protocols work the same way. Some prioritize accuracy while others prioritize speed, depending on the application's needs.

How the Internet Works (Simplified)

We're going to conclude this chapter with a simple web request example to illustrate how the Internet works. In this example, we're located in California. Our home LAN consists of two computers connected to a switch and a router. We want to visit InstructorAlton.com, whose web server is located across the country in a data center in Georgia. On one of the desktop computers within the LAN, we open a web browser, type https://www.instructoralton.com into the address bar, and press Enter, which initiates an HTTPS request. From our perspective, the website loads in our web browser in under a second, but in reality, a lot's happening. So, let's walk through this web request process:

1. Translation: Finding the Address

Your computer asks a DNS server, "What's the numeric address for InstructorAlton.com?" The DNS server responds with the IP address, something like 172.67.166.113. So now your computer knows where to send your request.

2. Local Network: Switch to Router

Your computer creates data packets containing your request and sends them to your home network switch. The switch examines the destination and recognizes it's outside your local network, so it forwards the packets to your router using MAC addresses.

3. Leaving Home: Your Router Takes Over

Your router examines the destination IP address and recognizes it's not on your local network. It consults its routing table and forwards your packets to your Internet Service Provider's network. This is where your data leaves your LAN and enters the Internet.

4. The Journey: Hopping Across Routers

Your request now travels across the Internet, hopping from router to router. Each router examines the destination IP address, looks up the best path in its routing table, and forwards your packets to the next router. Crucially, each router knows only the next best hop, not the entire path, much like a relay race where each runner knows only whom to pass the baton to next. Your data may pass through 10–20 routers, crossing multiple networks and ISPs, as it gradually makes its way from California to Georgia. The specific path depends on factors like network congestion, router availability, and connection speeds.

5. Arrival: Reaching the Destination

Eventually, your packets reach the network in Georgia, where the web server is located. The final router forwards them to a switch, which delivers them to the specific web server using its MAC address.

6. Response: The Return Journey

The web server processes your request, retrieves the webpage, and sends response packets back to you. These packets travel back across the Internet, hopping from router to router, but they might take a completely different path than your original request took. Network conditions change constantly, and routers continually select the best available path.

> ! **COMMON MISTAKE**: Many people assume data takes the same path in both directions. In reality, your request might go through Chicago while the response comes back through Denver. Routers choose the best path based on current conditions.

7. Home Again: Back to Your Computer

When the response packets reach your home router, it forwards them to your switch, which then forwards them to your computer. Your browser receives the webpage data and displays it on your screen, where this entire process typically happens in less than a second.

> 🎯 **KEY POINT**: This single webpage request involved DNS resolution, multiple protocol layers (HTTP, TCP, IP), two types of addressing (IP and MAC), two types of hardware (switches and routers), and potentially dozens of networks working together seamlessly.

Bringing It All Together

Computer networks are the primary battlefield of cybersecurity. Everything you've learned in this chapter, from protocols and addresses to switches and routers, represents both the infrastructure we depend on and the attack surface hackers target.

The distinction between physical and logical connections matters because attackers can exploit both. Understanding protocols is crucial because many attacks exploit weaknesses in how devices communicate, such as spoofing addresses, intercepting unencrypted traffic, or manipulating routing decisions.

The difference between LANs and WANs, MAC and IP addresses, and switches and routers isn't just academic—it defines how data flows and where vulnerabilities exist. Attackers understand this architecture intimately and exploit it at every layer.

When you typed a web address and watched a page load in under a second, you witnessed a complex choreography of protocols, addresses, switches, and routers working together across potentially dozens of networks. Understanding this process helps you appreciate both the elegance of network design and the many points where security can fail.

In later chapters, where we discuss and explore malware and network attacks, you'll see how attackers exploit these very concepts. The foundation you've built here is essential, as it'll make the threats we face and the defenses we build much clearer.

Key Takeaways

- **Networks Are Everywhere**: From your home Wi-Fi to global corporate systems, networks are the foundation of modern computing and communication.
- **Two Types of Addresses**: MAC addresses are physical (burned into the hardware), while IP addresses are logical (assigned by software). Both are essential for network communication.
- **Switches vs. Routers**: Switches connect devices within a LAN using MAC addresses, while routers connect different networks using IP addresses. Think of switches as organizing mail within a building and routers as directing mail between cities.
- **The TCP/IP Model**: The Internet operates on a four-layer model (Application, Transport, Internet, and Network Access) that works like the postal service. Each layer adds its own information (such as addressing and routing details) to help deliver your data.
- **Common Protocols Have Specific Jobs**: For example, HTTP/HTTPS for web browsing, FTP for file transfers, SSH for secure remote access, DNS for translating domain names, and DHCP for automatic IP address assignment.
- **The Big Picture**: A single web request involves multiple protocols, two types of addressing, various network devices, and potentially dozens of networks working together seamlessly. Understanding this process is essential for identifying security vulnerabilities.

Quick Action: See Your Network in Action

Networking concepts can feel abstract when you're reading about IP addresses, DNS, and routing protocols. The best way to make these concepts concrete is to see them working in real-time on your own computer. This quick action exercise will help you observe the networking fundamentals you just learned in practice, transforming theory into hands-on understanding.

Note: This quick action requires a desktop or laptop computer. If you only have a phone or tablet, you won't be able to complete these steps.

Try These Simple Commands

First, Open Your Command Line:

Windows: Click the Start menu or search button, type "command prompt" and click on Command Prompt when it appears.

Mac: Click the magnifying glass (Spotlight) in the top-right corner, type "terminal" and click on Terminal when it appears.

You'll see a window with text. This is where you'll type commands for steps 1 through 3 below.

Step 1: Find Your IP address

- In the command prompt or terminal window, type `ipconfig` (Windows) or `ifconfig` (Mac) and press Enter.
- Look for your local IP address in the results (usually starts with 192.168.x.x); you may have several.

Step 2: See DNS in Action

- In the command prompt or terminal window, type `nslookup google.com` and press Enter.
- Watch as DNS translates the domain name to an IP address.

Step 3: Trace a Route

- In the command prompt or terminal window, type `tracert google.com` (Windows) or `traceroute google.com` (Mac) and press Enter.
- Watch your data hop from router to router across the Internet.

These simple experiments make the abstract concepts concrete and show you the networking infrastructure working in real-time.

PART 1 CONCLUSION

Bringing It All Together

You've now completed the foundational portion of your cybersecurity education. You understand that cybersecurity isn't just a technical problem; it's a critical business and societal need driven by our increasingly connected world. You've learned that hackers follow systematic methodologies, whether they're criminals exploiting vulnerabilities for profit or ethical hackers testing defenses to make organizations stronger.

Most importantly, you now have the networking vocabulary that underpins every cybersecurity concept we'll explore. When we discuss defenses like firewalls, you'll understand what they do to protect network traffic. When we talk about man-in-the-middle attacks, you'll grasp how they exploit data flows between systems. When we examine DNS spoofing, you'll know exactly what DNS does and why attackers target it.

This solid foundation is what prepares you for what comes next: the core security principles that guide *how* we think about and implement cybersecurity defenses. Your expertise is already growing. Let's keep building.

PART 2: CORE SECURITY PRINCIPLES

Overview

Now that you understand the threat landscape and networking fundamentals, it's time to learn the defensive principles that guide cybersecurity practice. Chapter 4 introduces the conceptual frameworks that security professionals use to make decisions, evaluate risks, and design protective measures.

You'll learn the fundamental concepts that answer critical questions: What are we trying to protect? What are the primary threats to those protections? Why do we need multiple layers of security instead of just one strong defense? How do we control who has access to what? How do we prove that someone took a specific action? And how do we decide where to focus our limited security resources?

Think of this chapter as your security philosophy toolkit. These principles will shape your approach to every security decision you make, from choosing which controls to implement to determining how much security is truly "enough" for your organization.

Chapter 4. Core Cybersecurity Principles

Introduction

So far, we've focused on understanding the threat landscape. You've learned what cybersecurity is and why it matters, how hackers think and operate, and the networking fundamentals that form the primary attack surface for cyberattacks. Before diving into specific threats in later chapters, you need to understand the core defensive principles that guide how we think about security.

This chapter will be your primer on the fundamental concepts you'll see applied throughout the rest of the book: the CIA and DAD triads, defense-in-depth, least privilege, AAA, non-repudiation, and risk management basics. With these principles in mind, you'll better understand the threats we explore and the defenses we implement. Let's get into it.

The CIA & DAD Triads

When we talk about information security, understanding what we're protecting and what we're defending against is essential. That's where two complementary frameworks come into play: the CIA triad (Confidentiality, Integrity, and Availability), which defines our security objectives, and the DAD triad (Disclosure, Alteration, and Destruction/Denial), which identifies the primary threats to those objectives. Think of it this way: the CIA triad represents your goals (what you want to achieve with your security program), whereas the DAD triad represents obstacles (the threats working against those goals). Together, these frameworks provide a comprehensive view of the information security landscape.

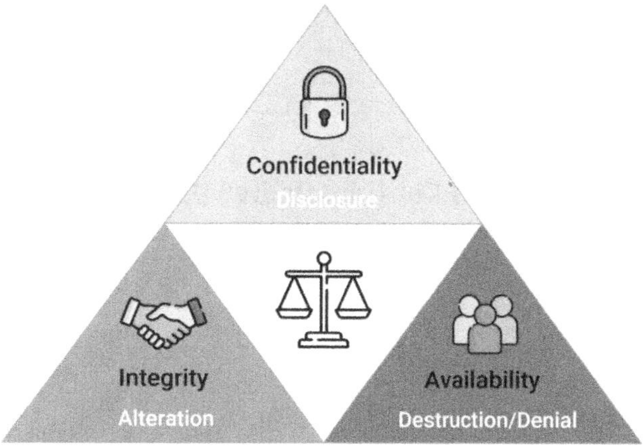

Figure 4.1: The CIA and DAD Triads

Confidentiality & Disclosure

Confidentiality is exactly what the name implies: ensuring that the data on our network and IT systems is secure and confidential. In other words, it's designed to prevent unauthorized disclosure of information on our network and IT systems. Confidentiality ensures that sensitive information remains accessible only to authorized individuals or systems. When confidentiality is compromised, unauthorized parties gain access to data they shouldn't see, potentially leading to identity theft, a competitive disadvantage, or regulatory violations, to name just a few potential outcomes.

Common Confidentiality Controls Include:

- **Authentication**: Implementing authentication measures that require people to log in before accessing data and verifying their identity before granting access.
- **Access Control**: Implementing physical and logical access control measures to sensitive areas and assets, such as data centers and file servers, ensuring only authorized personnel can reach critical resources.
- **Encryption**: Utilizing encryption to store and transmit sensitive information securely, rendering data unreadable to anyone without the proper decryption key.

> **EXAMPLE: Capitol One Breach (2019)**
>
> In 2019, Capital One discovered that a former AWS employee exploited a misconfigured web application firewall to access over 100 million customer records. The breach underscored the importance of implementing robust access controls and encryption for sensitive financial data stored in cloud environments. This incident illustrates how a single misconfiguration can compromise the confidentiality of millions of people.

What Threatens Confidentiality? Disclosure

Disclosure is the unauthorized access to or exposure of sensitive information; it's what happens when confidentiality fails. When disclosure occurs, private information becomes available to unauthorized individuals, whether through deliberate attacks, accidental exposure, or system misconfiguration.

Common Disclosure Threats Include:

- **Data Breaches**: Attackers exploit vulnerabilities to access sensitive data, such as customer information, financial records, or intellectual property, and then exfiltrate it for malicious purposes.
- **Man-in-the-Middle Attacks**: Attackers intercept communication between two computers to eavesdrop on sensitive information, such as login credentials, financial data, or private conversations.

- **Weak Access Controls**: Poorly implemented access controls, such as shared passwords, a lack of encryption, or overly permissive file permissions, which can lead to unauthorized disclosure.

For example, if a hacker exploits a web application vulnerability to access a database containing unencrypted customer credit card information, then sensitive financial data is disclosed to unauthorized parties, compromising confidentiality that encryption and stricter access controls are designed to protect.

Integrity & Alteration

Next up is integrity, which is all about ensuring that we have measures in place to prevent unauthorized manipulation of our data. In other words, making sure that people who shouldn't have access to our data aren't modifying, tampering with, or corrupting it. Data integrity means your information remains accurate, consistent, and unaltered unless authorized changes are made. Without integrity controls, data can be modified by attackers, corrupted by system errors, or altered by unauthorized users, all of which can lead to issues such as incorrect business decisions, financial losses, or compliance violations.

Common Integrity Controls Include:

- **Data Validation**: Implementing data validation to combat buffer overflow, SQL injection, and other data integrity attacks by verifying that input data meets expected formats and values.
- **Logging**: Logging all activities, both physical and logical, as a means to audit IT systems and data integrity, creating an accountability trail that helps detect unauthorized modifications.
- **Backup and Recovery**: Using backup procedures as a way to recover from data tampering and corruption, ensuring you can restore accurate data when you need to.

EXAMPLE: CCleaner Supply Chain Attack (2017)

In 2017, hackers compromised the update mechanism for CCleaner, a popular PC optimization tool, and distributed malware to over 2.27 million users through a legitimate-looking software update. This attack demonstrated the critical importance of verifying the integrity of software updates and downloads. Users trusted that the update was genuine, but the integrity of the software had been compromised.

What Threatens Integrity? Alteration

Alteration is the unauthorized modification of or tampering with data, and it occurs when integrity fails. When alteration occurs, information becomes inaccurate, corrupted, or manipulated, potentially leading to incorrect business decisions, financial fraud, or system malfunctions.

Common Alteration Threats Include:

- **SQL Injection Attacks**: Attackers inject malicious SQL code into web applications to modify or delete database data, thereby manipulating information for fraudulent purposes or sabotage.
- **Malware**: Certain types, such as ransomware, can alter or encrypt files, rendering them inaccessible or corrupted until a ransom is paid.
- **Insider Threats**: Employees or contractors with malicious intent may alter critical data, such as financial records, system configurations, or product specifications, either for personal gain or to harm the organization.

For example, if a hacker successfully uses an SQL injection attack to modify account balances in a financial database, transferring funds to their own account, then the integrity of the financial data has been compromised through unauthorized alteration, turning bad data into direct fraud.

Availability & Destruction/Denial

Finally, availability ensures that our IT servers, systems, and data are accessible when the business needs them. For example, if we're a business that operates from 6:00 AM to 6:00 PM, we want to make sure that everything is available during that timeframe to operate effectively and efficiently. Availability ensures that information, data, and IT systems remain accessible to authorized users when needed. Without availability, even perfectly secure data becomes useless. If your employees can't access the systems they need, then even the most advanced security becomes an obstacle to business operations.

Common Availability Controls Include:

- **Redundancy**: Using power, server, and data redundancies to ensure data remains available, even after an IT system failure, natural disaster, or cyberattack
- **Data Backups**: Using data backups to ensure we can quickly restore data when needed, minimizing downtime and business disruption
- **Balancing IT Security Controls**: Availability can be hampered by excessive security controls for confidentiality and integrity.

EXAMPLE: Dyn DNS DDoS Attack (2016)

In October 2016, attackers used a massive botnet of infected IoT devices, including cameras and routers, to launch a DDoS attack against Dyn's DNS service. By overwhelming Dyn's DNS servers with traffic, the attackers prevented users from resolving domain names for major websites like Twitter, Netflix, and Reddit. Although these sites were fully operational, users were unable to reach them because DNS queries failed. The attack disrupted access for several hours, demonstrating how availability attacks can cripple services by targeting critical infrastructure dependencies rather than the services themselves.

What Threatens Availability? Destruction & Denial

Destruction and denial represent disruptions to accessing our data, or a complete loss of data and systems; they're what happens when availability fails. This category encompasses temporary denial-of-service (making systems unavailable) and permanent destruction of data or infrastructure.

Common Destruction/Denial Threats Include:

- **Distributed Denial-of-Service (DDoS) Attacks**: Attackers flood servers with traffic, overwhelming them and making services unavailable to legitimate users, though typically without destroying data.
- **Ransomware**: Attackers install malware that encrypts data or locks systems, denying access until a ransom is paid. In some cases, ransomware can also destroy data if demands aren't met or if the attacker simply deletes encryption keys.
- **Data Wiping Malware**: This is malicious software specifically designed to permanently delete or corrupt data, such as wiper malware. It's used in destructive cyberattacks that aim to cause maximum damage rather than financial gain.
- **Physical Destruction**: Hardware failures, natural disasters (such as fires, floods, or earthquakes), or deliberate sabotage can result in the complete destruction of systems or data, potentially causing permanent loss.

For example, if a ransomware attack encrypts all files on a company's network and the attacker deletes the encryption keys when the ransom isn't paid, then the data is permanently destroyed, denying the availability of critical business information, which is lost forever without proper backups.

A Balancing Act

Implementing and using the CIA triad is a balancing act because you typically reduce availability whenever you implement confidentiality and integrity security measures. You want to secure your IT assets and data, especially against insider threats and cybercriminals, but only up to the point where you're not hindering operations too much.

Think About It This Way:

- **Too Much Security, Not Enough Availability**: If a bank made its ATM so secure that customers had to go through 10 authentication steps and wait 5 minutes for each transaction, nobody would use it. The system would be secure but useless.
- **Too Much Availability, Not Enough Security**: If that same bank removed all security to make transactions instant, anyone could access anyone else's account. This would be extremely fast, but completely insecure.
- **The Sweet Spot**: What banks actually do, which is implementing reasonable security measures, including PIN verification, card chip authentication, and transaction limits, that protect customers while still allowing quick and convenient access.

Organizations need to find the right balance for their specific needs. A hospital's medical records system might prioritize availability (doctors need immediate access in emergencies) while still maintaining strong confidentiality and integrity controls. On the other hand, a financial trading platform might prioritize integrity (ensuring transactions are accurate) and availability (trades must execute quickly) while implementing appropriate confidentiality measures.

IT security teams typically aim to lock everything down with maximum confidentiality and integrity controls, whereas business operations teams prioritize maximum availability to ensure employees can work efficiently. The solution lies in compromise: risk-based decisions made jointly by business and IT security leaders to achieve appropriate levels of confidentiality, integrity, and availability tailored to the organization's specific needs.

KEY POINT: Every security control you add makes systems slightly less convenient. The goal isn't maximum security; it's appropriate security that protects your assets without crippling your business operations.

COMMON MISTAKE: A common pitfall for business owners is implementing security controls without considering their impact on business operations. Security measures that prevent employees from working efficiently may cause financial harm to the business and create pressure to reduce security, ultimately making you less secure, not more.

Two Sides of the Same Coin

The relationship between the CIA and DAD triads is straightforward: when confidentiality fails, disclosure occurs. When integrity fails, alteration happens. When availability fails, destruction or denial results. The CIA triad represents our objectives, while the DAD triad represents the consequences when we fail to achieve those objectives.

Together, they provide a comprehensive view of information security:

- **CIA**: What we aim to protect (our goals)
- **DAD**: What we aim to defend against (our adversaries)

The goal of implementing CIA triad principles is specifically to mitigate DAD triad outcomes. Every security control you implement should address one or more aspects of the DAD triad—preventing unauthorized disclosure, unauthorized alteration, or destruction of data/denial of access.

> **KEY POINT**: Understanding both triads helps you think strategically about security. When evaluating a security control, ask yourself, "Which aspect of the CIA triad does this protect?" and "Which aspect of the DAD triad does this prevent?"

Understanding what to protect leads us to the next step: how to implement these protections in practice.

Defense in Depth

The next core cybersecurity principle we'll discuss is defense in depth, which is a layered approach to IT security. With defense in depth, you implement multiple types of security controls throughout your network and IT systems. By layering your defenses, you significantly reduce your risk and make it much harder for attackers to succeed.

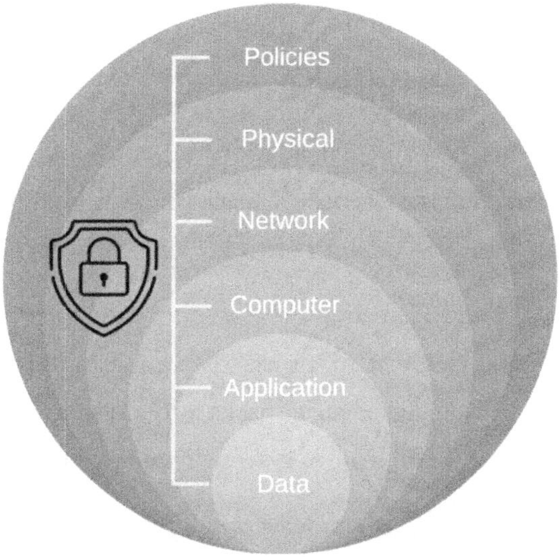

Figure 4.2: Defense in Depth

Defense in depth means securing every layer of your IT infrastructure. Rather than relying on a single security measure, you implement controls at multiple levels. This includes physical security, network perimeter defenses, internal network controls, endpoint protection, application security, data encryption, and so forth.

The entire premise of defense in depth is that by adding multiple layers, you make it much more difficult for hackers to breach your IT infrastructure. If you only had one security measure in place, say a perimeter firewall and nothing else, then only that single control would stand between a hacker and your data. Once they bypass that, they have free rein.

However, if you layer multiple different security measures throughout your infrastructure, you make it much harder for attackers to succeed. Even if they bypass the perimeter firewall, they still face additional security controls, including obstacles such as network segmentation, access controls, endpoint protection, encryption, and monitoring systems. As they penetrate each layer, their risk of detection increases, requiring additional time, effort, and sophistication.

Remember from Chapter 2 that hackers follow a systematic methodology. Defense in depth disrupts this methodology at multiple points throughout their attack. Even if attackers complete reconnaissance and identify vulnerabilities, they still face layers of protection during the exploitation stage. Each layer they penetrate forces them to repeat their attack process, which increases their risk of detection with every attempt. Defense in depth is a hallmark feature of effective cybersecurity. As we explore fundamental countermeasures in Chapter 8, you'll see how this layered approach comes into play across every aspect of security.

> **Key Point**: No single security control is perfect. Firewalls can be bypassed, anti-malware can be evaded, and passwords can be compromised. Defense in depth ensures that when one layer fails, others are still protecting your assets. It's about creating redundancy in your security posture.

Least Privilege

Another fundamental aspect of cybersecurity is the principle of least privilege, which forms the cornerstone of effective access control. Least privilege is all about limiting what people, systems, processes, and applications have access to on our network.

This is a fundamental cybersecurity principle because, when you think about it from a security perspective, giving a user, system, process, or application access to more resources than needed creates unnecessary risk. Let's say a malicious user hacks into a system and gains access to an account with unrestricted network access. In that scenario, you've essentially opened the door for them to access everything on your network.

When you implement least privilege, you limit what people have access to, which systems they can reach, and what applications they can use. It's about giving each entity exactly what it needs to function and nothing more. Let's examine some practical examples of implementing least privilege that everyone can relate to:

Personal Device Examples

You already practice least privilege in your daily life, even if you don't realize it. When you install a smartphone app, you grant only the permissions it needs. For example, a flashlight app doesn't need access to your contacts, so you shouldn't grant it, and if it asks, that's a red flag. Similarly, when using your computer, it's best to use a standard user account for daily tasks and reserve the administrator account for when you actually need elevated permissions.

Work Environment Examples

In work environments, least privilege takes many forms. Marketing employees are granted access to marketing files but not financial records. Additionally, employees who don't work after hours have their building and network access restricted to working hours. Furthermore, a company's public-facing website is designed so it can only read product information from databases, not modify product descriptions or prices.

The Common Thread

The common thread is simple: we should grant users, applications, and systems only the permissions they need to function, and nothing more. When something gets compromised, limited access means limited damage. Let's examine a real-world example of what happens when least privilege isn't properly implemented:

EXAMPLE: Twitter Hack (2020)

In 2020, Twitter experienced a major security incident when attackers used social engineering to trick employees into providing access to accounts with elevated administrative privileges. The attackers then accessed internal tools to hijack high-profile accounts, including those of Barack Obama, Elon Musk, and Bill Gates, in order to promote a Bitcoin scam. Stricter least privilege controls on administrative access could have significantly limited the damage. If those employee accounts had only the permissions necessary for their specific job functions, rather than broad administrative access, the attackers wouldn't have been able to compromise such high-profile accounts. We'll explore this attack in detail in Chapter 7's case study, where you'll see how voice phishing bypassed technical security controls and why the human element remains the weakest link.

AAA: Authentication, Authorization, & Accounting

Building on the CIA and DAD triads, defense-in-depth, and least privilege, let's explore another core information security framework: AAA, commonly pronounced as "triple A." This framework—authentication, authorization, and accounting—provides the foundation for controlling access to data and system resources, enforcing policies, and auditing actions.

Additionally, AAA works hand in hand with the principles we've already discussed, particularly least privilege. Think of it this way: least privilege defines the "what" (what access people should have), and AAA provides the "how" (how we verify identity, enforce access limitations, and track what people do with that access).

Authentication

Authentication is the process of verifying a user's identity, proving that you are who you claim to be. This happens every time you log into a system. You're already familiar with the process, even if you've never heard of the technical term. You authenticate constantly in your daily life. When you log in to Facebook, YouTube, Twitter, or your online bank with your username and password, the system verifies your credentials. When you unlock your smartphone with your fingerprint, the device confirms you're the authorized owner. These are examples of authentication, proving you are who you claim to be.

Authentication answers the question, "Who are you?" It's the first step in access control because before a system can decide what you're allowed to do, it needs to know who you are. Without authentication, there's no way to enforce authorization or maintain accountability.

Common authentication methods include:

- **Something You Know**: Passwords, PINs, security questions
- **Something You Have**: Smart cards, security tokens, mobile phones for one-time codes
- **Something You Are**: Biometrics like fingerprints, facial recognition, or iris scans

We'll explore multi-factor authentication, which combines two or more of these methods, in detail in Chapter 8 when we discuss fundamental threat countermeasures.

Authorization (Access Control)

Once you've authenticated and the system knows who you are, authorization determines what you're allowed to do and which resources you can access. Authorization is the implementation of access control policies. More specifically, it's how we enforce least privilege in practice. Think about it this way: authentication confirms your identity, while authorization determines your access permissions. Two people might both successfully authenticate to the same system, but they'll have very different capabilities based on their authorization levels.

For instance, if I log in to a Windows system as an administrator and you log in as a standard user, we've both successfully authenticated. However, my authorization permits me to install and uninstall applications, change system settings, and access all system files. Your authorization, on the other hand, restricts you to running approved applications and accessing only your own files. The system authenticated both of us, but our authorization levels differ dramatically.

Furthermore, authorization also implements role-based and department-based access control. If you're assigned to the human resources department, your authorization should grant access only to HR data and systems, not to the accounting department's financial records or the marketing department's marketing records. This separation ensures that individuals can access the resources they need for their jobs while protecting sensitive information from unauthorized access.

Accounting

The third component of AAA is accounting, which works in conjunction with auditing and oversight. Accounting's purpose is to track and record user actions and system access, essentially creating a detailed log of who did what, when, and where.

This audit trail serves multiple critical security functions:

- **Creates Accountability**: When people know their actions are logged, they're more careful and responsible.
- **Supports Incident Investigations**: You can trace the sequence of events, identify which accounts were involved, and understand the scope of a breach.
- **Ensures Compliance**: Many regulations require detailed audit logs of access to sensitive data.
- **Supports Integrity**: If someone tampers with critical files, the logs show who accessed them and when.

Examples of accounting in practice include logging user authentication events (successful and failed login attempts), recording file access and modifications, tracking privileged actions such as system configuration changes, and monitoring network connections and data transfers.

> **Key Point**: Authentication, authorization, and accounting work together as a complete access control system. Authentication verifies identity, authorization defines permissions based on that identity, and accounting tracks actions to maintain accountability and support investigations. All three components must function together for effective security.

Non-Repudiation

The "Accounting" principle of AAA naturally leads to another core concept: non-repudiation. While accounting tracks what happened, non-repudiation provides undeniable proof of who did it, preventing someone from credibly denying they took an action. Non-repudiation is a concept and methodology used to prevent an entity from denying that an action took place. An "entity" could be a user, a device on the network, or any other person or system in your IT environment. From the perspective of someone on your network, such as an employee, you use non-repudiation to ensure that you can track their actions and have a definitive method to confirm whether they performed a given action or not.

Think of non-repudiation as creating undeniable proof of actions. It's one thing to log that someone accessed a file (accounting), but non-repudiation provides cryptographic or procedural proof that a specific person performed a specific action, making it impossible for them to credibly deny it later.

Digital Signatures

The most common example of non-repudiation is digitally signed documents. When you digitally sign a PDF document or an email, you're creating cryptographic proof that you, and only you, took that action. This signature is tied to your unique digital identity (your digital certificate) and cannot be forged or repudiated.

Many organizations configure their email servers to automatically add digital signatures to outgoing emails on behalf of users when they're logged in to their accounts. This provides non-repudiation for email communications, allowing the recipient to verify that the message genuinely came from the claimed sender and preventing the sender from later claiming they didn't send it.

Sometimes you'll want to create digital signatures intentionally to prove your approval or agreement with something. Furthermore, organizations also commonly enforce digital signatures automatically, ensuring that employees can't later claim "that wasn't me" when confronted with their actions.

Audit Logging

While accounting tracks what happened (as we discussed in AAA), combining comprehensive audit logs with strong authentication creates non-repudiation. If the logs show that your account accessed a sensitive file at 2:47 PM on Tuesday, and the system required authentication to access that file, it becomes very difficult for you to credibly deny that you accessed it. The combination of authentication logs, access logs, and proper security controls creates a chain of evidence that establishes non-repudiation; you can't claim "someone else used my account" when multi-factor authentication was required.

Physical Security Non-Repudiation

Non-repudiation also applies to physical security. For example, if entering your office building requires showing an employee ID badge and signing in with a security guard, that sign-in record provides non-repudiation for physical access. Furthermore, security camera footage, combined with badge access logs, provides another form of physical non-repudiation; there's video evidence of who entered secure areas and when, making it impossible to deny physical presence. This becomes particularly important when investigating physical security incidents or when restricted areas contain sensitive equipment or information.

🎯 **Key Point**: Non-repudiation creates undeniable proof of actions through cryptographic signatures, audit logs, or physical records. It's the security principle that says "we can prove you did this, and you can't credibly deny it." This serves both as a deterrent (people are more careful when they know their actions can be proven) and as evidence during investigations.

📘 **TECH TALK: Digital Certificates**

Digital signatures rely on digital certificates, which are electronic credentials that bind a digital identity to a person, device, or organization. Think of a digital certificate as a digital ID card that contains your public encryption key along with information about who you are. When you digitally sign something, you're using your private key (which only you possess) to create a signature that others can verify using your public key (contained in your digital certificate). This cryptographic process ensures that the signature could only have been created by someone possessing your private key, providing mathematical proof of non-repudiation.

Risk Management Basics

We're going to conclude this chapter with a brief introduction to the principle that underlies all security decisions: risk management. At its core, cybersecurity is the practice of identifying potential threats and making informed decisions about how to handle them. We'll explore formal risk management processes in detail in Chapter 10, when we discuss security management and culture. However, for now, it's important you understand this basic concept, as it's the foundation for all the security principles we'll discuss in this book.

In cybersecurity, risk refers to the possibility that something bad will happen to your systems or data. Furthermore, risk increases when two things come together:

- **Vulnerability**: A weakness in your defenses
- **Threat**: Someone or something that could exploit the vulnerability

For example, if you use a weak password (vulnerability) and hackers are actively trying to break into your account (threat), you face a significant risk. However, if either factor changes, such as when you create a strong password or hackers begin targeting someone else, your level of risk decreases accordingly.

Think about it this way: Leaving your garage door open all day isn't automatically a problem in itself. The risk comes from the combination of your open garage door (vulnerability) and the possibility that a burglar might try to steal something from within it (threat). If you keep your garage door closed or move to a neighborhood with low crime rates, your risk drops significantly. In other words, remove either the vulnerability or the threat, and you've reduced your risk.

Key Point: Risk isn't just about vulnerabilities or just about threats; it's about the intersection of the two. You can reduce risk by addressing either side of the equation. Sometimes it's easier to fix vulnerabilities, whereas other times it's more feasible to reduce exposure to threats.

How Organizations Manage Risk

Organizations handle cybersecurity risks in four main ways, and understanding these options will help you think strategically about security:

- **Risk Acceptance:** Sometimes the risk is so small, or the cost of fixing it is so high, that organizations simply decide to live with it. A small business might accept the risk of not having a backup generator because power outages are rare in its area, and purchasing one is cost-prohibitive. This is a conscious decision to acknowledge the risk exists, but not to take action to reduce it.
- **Risk Avoidance:** A business can avoid risk by completely stopping a risky activity. If cloud storage seems too risky for highly sensitive data, a business might choose to keep everything on-premises instead. If a particular business process creates security risks that can't be adequately mitigated, then the business may discontinue that process altogether. Risk avoidance is the most comprehensive way to eliminate risk, but it often means forfeiting business capabilities.
- **Risk Mitigation:** This is what organizations most often do: implement security controls to reduce the likelihood that something bad will happen or limit the impact if it does. Installing anti-malware software, using firewalls, training employees, and implementing access controls are all examples of reducing risk through risk mitigation. The organization doesn't eliminate the risk entirely (that's usually impossible), but it brings it down to an acceptable level.

- **Risk Transference:** To transfer risk, a business shifts the financial consequences of a risk to someone else, typically through insurance. While the risk still exists, if something bad happens, the insurance company pays for the damages. This doesn't reduce the likelihood or impact of security incidents, but it transfers the financial burden to a third party. It's common practice to use insurance as one component of an overall risk management strategy.

Every security control we've discussed in this chapter, the CIA triad, defense-in-depth, least privilege, AAA, and non-repudiation, is ultimately about reducing risk to an acceptable level. We'll explore how organizations formally manage risks in Chapter 10 when we discuss risk management in much more detail. For now, simply understand that cybersecurity decision-making is fundamentally about identifying risks and determining how to address them.

> **DEFINITION: Risk Management** is the process of identifying, assessing, and responding to risks by accepting, avoiding, reducing, or transferring them to an acceptable level that balances security with business operations.

Bringing It All Together

This chapter introduced the defensive principles that guide cybersecurity practice. These principles work together as an integrated framework for protecting systems and data. The CIA and DAD triads provide a complete picture of information security. CIA defines what we protect: confidentiality, integrity, and availability, while DAD identifies what threatens us: disclosure, alteration, and destruction/denial. Every security control serves CIA objectives while preventing DAD outcomes.

Defense in depth recognizes that no single control is perfect. By layering multiple defenses, you force attackers to overcome multiple barriers, increasing their likelihood of detection and failure at each stage. Least privilege limits damage by ensuring compromised accounts have minimal access. The AAA framework provides the practical mechanisms for implementing these principles: verifying identity, enforcing permissions, and tracking actions. Furthermore, non-repudiation creates undeniable proof of actions, supporting both deterrence and investigations.

Finally, risk management ties everything together. You can't eliminate all risk, so you identify threats and vulnerabilities, assess their likelihood and impact, and respond by accepting, avoiding, mitigating, or transferring risk to acceptable levels. This ensures you invest security resources where they'll have the greatest impact. These principles—the CIA and DAD triads, defense in depth, least privilege, AAA, non-repudiation, and risk management—form the foundation for everything that follows. As we move into specific countermeasures in Chapter 8, you'll see these principles applied through concrete security controls and technologies.

Key Takeaways

- **The CIA and DAD triads work together**: CIA defines what we protect (confidentiality, integrity, and availability) while DAD identifies what threatens us (disclosure, alteration, and destruction/denial).
- **Defense-in-depth layers multiple security controls** throughout your infrastructure, forcing attackers to defeat multiple defenses rather than just one.
- **Least privilege limits access** to only what's necessary: users, systems, and applications get exactly the amount of access they need to function and nothing more.
- **The AAA framework** (Authentication, Authorization, and Accounting) provides the practical mechanisms for controlling access, verifying who you are, defining what you can do, and tracking what you do.
- **Non-repudiation creates undeniable proof of actions** through digital signatures, audit logs, and physical records, thereby preventing individuals from denying their actions.
- **Risk management involves identifying and responding to risks** by accepting, avoiding, mitigating, or transferring them to acceptable levels that balance security with business operations.

🔧 Quick Action: Assess Your Personal Security Posture

You've just learned about several fundamental principles that guide all cybersecurity decisions. Before you can improve your security, you need to understand where you currently stand. This quick assessment helps you identify your strengths and gaps using the concepts from this chapter, giving you a clear picture of what needs attention first. So in this quick action exercise, you'll work through a short checklist based on concepts discussed in this chapter to help identify which strengths and weaknesses to focus on.

Check Your Current Security

CIA Triad Check:

- **Confidentiality:** Do you use strong, unique passwords for important accounts?
- **Integrity:** Do you have backups of important files?
- **Availability:** Can you access your important data when you need it?

Defense in Depth Check:

- Do you have anti-malware software installed and up to date?
- Do you have multi-factor authentication enabled on important accounts?
- Do you review login activity on important accounts?

Least Privilege Check:

- Do you use a standard user account for daily tasks (not an administrator account)?
- Do you grant apps only the permissions they actually need?

Identify Your Biggest Gap:

- Based on the checks above, what's your most significant vulnerability?
- Write down one specific action you'll take this week to address it.

This self-assessment is the foundation of personal cybersecurity awareness and protection. Now that you understand where you currently stand, you can focus your efforts on the areas that need the most attention. Keep this in mind as you progress through the book and work on future Quick Actions, building stronger security one step at a time.

PART 2 CONCLUSION

Bringing It All Together

You've now learned the fundamental principles that guide all cybersecurity practice. The CIA triad showed you the three core goals of information security: keeping data confidential, maintaining its integrity, and ensuring it's available when needed. The DAD triad revealed the primary threats to those goals: attempts at disclosure, alteration, destruction, or denial. Defense-in-Depth taught you the critical lesson of layered defenses—why relying on a single control is dangerous. Least privilege, AAA, and non-repudiation gave you the concepts for controlling access and maintaining accountability. And risk management provided a framework for making intelligent decisions about where to focus limited resources.

These principles aren't just abstract theory—they're the foundation you'll use to evaluate every security decision. When you read about a breach in the news, you'll understand which principles were violated. When you implement security controls in the coming chapters, you'll see how each one supports these core concepts.

With this philosophical foundation in place, you're ready to dive deep into the specific threats you'll face. Understanding these principles first ensures you'll see not just what the threats are, but why they're dangerous and how they violate the fundamental goals of security. It's time to explore the threat landscape.

PART 3: UNDERSTANDING THE THREATS

Overview

Armed with foundational knowledge and core security principles, you're ready to face the adversary head-on. This section takes you deep into the diverse and constantly evolving threat landscape that security professionals defend against daily. We'll begin with malware in Chapter 5: malicious software designed to damage systems, steal data, or gain unauthorized access. Chapter 6 expands your threat awareness beyond malware to include attacks that exploit network protocols, web applications, and emerging technologies like artificial intelligence. Finally, Chapter 7 addresses what many consider the most dangerous vulnerability of all: human psychology—exploring how attackers manipulate people into bypassing security controls through deception and social engineering. By the end of Part 3, you'll understand:

- The major categories of malware and how each operates
- Common network and application-layer attacks
- How attackers exploit human trust and psychology
- Why technical controls alone can never provide complete protection
- The real-world impact of these threats through detailed case studies

This knowledge transforms you from someone who knows security exists to a practitioner who truly understands what you're up against and how to categorize the threats.

Chapter 5. Malware Threats

Introduction

In this chapter, we'll explore malware, which is malicious software designed to damage systems, steal data, or gain unauthorized access to a victim's network or devices. This includes viruses, worms, Trojan horses, logic bombs, spyware, rootkits, and ransomware, as well as how hackers deploy them. Let's get into it.

Viruses

Let's begin with viruses, the most well-known type of malware. In the early days of cybersecurity, anti-malware software was simply called "anti-virus" because viruses were the most prevalent threat. While we now face a much broader landscape of malicious software, understanding how viruses work is still fundamental.

A virus is a piece of malicious code that attaches itself to a legitimate program or file. It remains dormant until a user takes a specific action that triggers it. For example, a macro-based virus hidden in an Excel spreadsheet will activate only when you open the file and enable macros. Once triggered, the virus executes its malicious objective, which could be anything from corrupting your operating system to installing spyware to stealing sensitive data.

What You'd Notice: Telltale signs of a virus infection include a sudden slowdown in computer performance, frequent program crashes, corrupted files that won't open, or an increase in unexpected pop-up windows.

How to Prevent: Since viruses rely on user interaction, the primary defense is caution. Avoid opening email attachments from unknown senders, keep your anti-malware software up to date to recognize known threats, and be extremely wary of enabling macros in documents. While anti-malware software is essential, it can't catch everything, especially new viruses. Your awareness and vigilance are key to defending against viruses.

> **KEY POINT**: Viruses can't execute on their own; they require a human action, such as opening a file or clicking a link to be triggered.

EXAMPLE: The ILOVEYOU Virus (2000)

The ILOVEYOU virus, also known as the Love Bug, was released on May 4, 2000, and spread to over 45 million computers within days. It arrived as an email with a fake love letter attachment that looked like a harmless text file. When opened, the virus unleashed its payload, sending copies of itself to every contact in the victim's email address book. The total cost of remediation was approximately $15 billion.

Worms

In stark contrast to viruses, worms are autonomous threats that don't require human interaction to spread. Don't assume you're safe from worms just because you didn't click on anything suspicious. Worms exploit technical vulnerabilities in your systems and network, and don't need you to click anything to do their damage. Once a worm gains a foothold on a single computer, often through an unpatched software vulnerability, it can replicate and travel across the network on its own. A worm actively scans the network for other vulnerable systems and infects them, creating hundreds or even thousands of copies of itself. This self-propagating nature is what allows worms to cause massive, global outbreaks in a matter of hours or days.

What You'd Notice: The most common symptom of a worm infestation is a dramatic slowdown of your network. As thousands of copies of the worm travel between devices, they consume massive amounts of bandwidth, causing websites to load slowly and file transfers to time out.

How to Prevent: Because worms exploit software vulnerabilities, the single most important defense is timely patch management. Keeping your operating system and applications updated closes the security holes that worms use to spread. Additionally, using network segmentation can contain a worm to one part of the network, preventing it from infecting the entire organization.

KEY POINT: A worm's defining characteristic is its autonomy; once on a network, it spreads without any further help. This is what makes worms so dangerous and capable of causing global outbreaks.

EXAMPLE: The Conficker Worm (2008)

The Conficker worm, first spotted in 2008, targeted Microsoft Windows by exploiting a known vulnerability and using dictionary attacks on administrator passwords to propagate through networks. It infected roughly 15 million computers across 190 countries, compromising systems at the French Navy and the United Kingdom Ministry of Defense, and ultimately cost more than $9 billion to remediate.

Trojan Horses

While viruses and worms often announce their presence through poor performance, our next threat is far more deceptive. A Trojan horse, true to its name, is malware that disguises itself as something you want. The term originates from the ancient Greek story where a giant wooden horse, presented as a gift to the city of Troy, secretly housed soldiers who emerged to conquer the city.

In cybersecurity, a Trojan horse is a program that appears to be a legitimate tool, such as a "free anti-virus scanner" or "PC cleanup utility," but contains hidden malicious code. When you install the seemingly useful software, it also secretly installs malware, such as spyware or ransomware, on your system.

What You'd Notice: You might not notice anything at first—that's the point. The software appears to work as advertised while silently executing its malicious purpose in the background. Eventually, you might see symptoms of the hidden malware, but the Trojan itself can remain undetected indefinitely.

How to Prevent: The most critical defense is source verification. Only download software from official sources like the Microsoft Store, Apple App Store, or the developer's official website. Be extremely skeptical of "free" versions of paid software, and don't trust software just because it's a professional-looking website or has positive reviews on random forums. Attackers are adept at creating convincing fake websites and reviews to lure victims. When available, verify file hashes to ensure the download hasn't been tampered with. A file hash is a unique digital fingerprint generated from a file's contents. Even a tiny change to the file, such as changing a single character in the code, will completely alter its hash value, revealing that the file is no longer authentic.

KEY POINT: If an offer seems too good to be true, like free premium software or a miracle computer cleaner, it probably is. Trojan horses prey on your desire to get something for nothing.

EXAMPLE: Merck Pharmaceuticals Trojan Horse Attack (2017)

In 2017, attackers used a Trojan horse to gain access to Merck's corporate workstations. Once inside, they moved through the network and caused outages across office and manufacturing systems. Production lines and scheduling were disrupted, some vaccine and medicine manufacturing was delayed, and supply and distribution were affected. Merck activated contingency plans, and losses were reported in the hundreds of millions, showing how a trusted-looking download can ripple into real-world medication delays and shortages.

Logic Bombs

Some malware isn't designed to have an immediate impact, but rather to detonate at a later time. A logic bomb is malicious code embedded in a program that executes only when specific conditions are met. This trigger could be a specific date, a particular system action, or the absence of a user account (e.g., after an employee has been terminated). While often planted by malicious insiders, an external attacker who's gained access could also deploy a logic bomb to cause future damage. The payload can be devastating, from deleting critical databases to crashing entire servers.

What You'd Notice: Absolutely nothing, until the bomb detonates. At that point, files may suddenly disappear, systems may crash, or data may become corrupted, with no obvious cause.

How to Prevent: Defending against logic bombs requires strong administrative and procedural controls. This includes implementing mandatory code reviews for changes to critical scripts, enforcing separation of duties, maintaining detailed audit logs, and having a process for immediately terminating all access for departing employees.

KEY POINT: Logic bombs are particularly dangerous because they can be planted by anyone with sufficient access—including disgruntled employees—and set to trigger long after the individual has left, making attribution difficult.

EXAMPLE: Union Bank of Switzerland (2002)

In 2002, a disgruntled IT administrator at Union Bank of Switzerland, Roger Duronio, planted a logic bomb after becoming unhappy with his annual bonus. The bomb was designed to delete critical files in an attempt to drive down the company's stock price. He was ultimately convicted, sentenced to 8 years in prison, and ordered to pay $3.1 million in restitution.

! **COMMON MISTAKE**: Having a poor off-boarding process. While preventative controls like code reviews are meant to stop a logic bomb from being planted, failing to immediately revoke all system access upon an employee's departure is a critical error. An active account could allow a disgruntled ex-employee to log back in to cover their tracks, cause new damage, or steal data after they've left.

Spyware

Not all malware is designed to be destructive. Some types, like spyware, are designed to be silent parasites, stealing information without you ever noticing. Spyware is designed to monitor your activities and collect information without your knowledge or consent, often installed via Trojan horses and running silently in the background. Once active, it can log your keystrokes, take screenshots, and capture sensitive data such as:

- **Keystrokes**: Keylogger software monitoring everything you type
- **Screenshots**: Taking periodic screenshots of your computer
- **Authentication Credentials**: Capturing your usernames and passwords
- **Personally Identifiable Information (PII)**: Collecting information to steal your identity
- **Web Form Data**: Monitoring and collecting data you enter into websites

What You'd Notice: Often nothing, as spyware is designed to be invisible. You might notice your computer running slightly slower, or you may only discover the breach after finding unauthorized purchases or fraudulent account activity.

How to Prevent: The best defenses include using reputable anti-malware software with real-time protection, downloading applications only from trusted sources, keeping your systems updated, and regularly monitoring your financial and online accounts for suspicious activity. Even pre-installed programs ("bloatware") can pose security risks, so review what's on your new devices and remove anything unnecessary.

KEY POINT: Spyware's goal is to remain undetected for as long as possible. The longer it runs, the more information it collects, making it critical to be vigilant about the software you install.

EXAMPLE: Lenovo Superfish (2014)

In 2014, Lenovo was found to have pre-installed Superfish adware on consumer laptops. The software acted as spyware by intercepting encrypted web traffic to inject advertisements, creating a serious security vulnerability that could allow attackers to steal passwords and financial data. Lenovo faced multiple lawsuits and was required to pay $3.5 million in settlements.

Adware

A less dangerous but more annoying cousin of spyware is adware, software designed primarily to display unwanted advertisements and collect marketing data about your browsing habits. Like spyware, it's often installed without clear consent, bundled with other free software. While some adware is relatively benign, it's considered malware because it's often installed deceptively and can be designed to act as spyware, tracking your browsing history to serve targeted ads or sell your data. While adware is often more of a nuisance than a direct threat, it can severely degrade your computer's performance and act as a gateway for more malicious types of malware.

What You'd Notice: The most obvious signs are constant pop-up advertisements, your browser's homepage changing without your permission, new and unfamiliar toolbars appearing, or your browser suddenly redirecting to advertising websites.

How to Prevent: Prevention focuses on safe browsing and installation habits. Use reputable ad blockers, avoid clicking on suspicious pop-up ads, only download software from official sources, and use anti-malware software that specifically detects adware. Many legitimate programs bundle adware, but you can usually opt out if you pay attention during installation—avoid clicking "agree" or "next" repeatedly without reading the prompts.

Ransomware

Ransomware is one of the most feared and financially damaging types of malware. It encrypts your personal files or your entire system, holding your data hostage until you pay a ransom, which can range from hundreds to millions of dollars. Attackers demand payment, usually in untraceable cryptocurrency like Bitcoin, in exchange for a decryption key. They often add pressure by threatening to permanently delete or publicly release your data if the ransom isn't paid within a tight deadline.

What You'd Notice: You're completely locked out of your files or your computer, and your screen displays a ransom note demanding payment and providing instructions, often with a countdown timer.

How to Prevent: The single most effective defense against ransomware is to maintain regular, offline backups of your important data. If your system is infected, you can restore it from a clean backup instead of paying the ransom. Other critical measures include keeping systems patched, using anti-malware software, and training your system's users to recognize phishing emails.

KEY POINT: Ransomware is one of the most financially damaging types of malware today. The best defense is having current, tested backups stored offline or in isolated locations where ransomware cannot reach them.

! **COMMON MISTAKE**: Assuming that paying the ransom guarantees you'll get your data back. Many victims pay and never receive a working decryption key, losing both their money and their data. This is why backups are essential.

Case Study: The WannaCry Ransomware Attack

To truly understand the devastating impact ransomware can have on a global scale, let's examine one of the most significant cyberattacks in history: the WannaCry ransomware attack of 2017.

Background

On May 12, 2017, the WannaCry ransomware attack spread like wildfire across the globe, infecting over 200,000 computers in more than 150 countries within days. The attack targeted computers running Microsoft Windows, encrypting user data and demanding ransom payments of $300 to $600 in Bitcoin. WannaCry caused widespread disruption to businesses, hospitals, and government agencies worldwide, highlighting critical vulnerabilities in outdated systems and poor patch management practices.

The Attack

WannaCry exploited a serious security flaw in Microsoft Windows known as EternalBlue, which targeted a vulnerability in the Server Message Block (SMB) protocol. Microsoft had released a fix for this flaw in March 2017, 2 months before the attack, but many organizations had failed to install the update. The attack spread through phishing emails and infected websites, but its worm-like capability is what made it so devastating. Once it infected one computer, it automatically spread to other unpatched computers on the same network without any human help.

Impact

The UK's National Health Service (NHS) was one of the hardest-hit organizations, with over 70,000 devices affected, including MRI scanners and blood-storage refrigerators. The attack disrupted hospital operations, forced the cancellation of appointments, and put patient care at risk. Globally, major organizations like FedEx and Renault were also affected. The total financial impact was estimated in the billions of dollars from lost productivity and recovery costs.

Lessons Learned

- **Patch Management is Critical**: A patch was available for 2 months before the attack. Regular, timely patching is one of the most effective defenses against known exploits.
- **Network Segmentation Limits Spread**: WannaCry spread rapidly within flat networks. Proper network segmentation could have potentially contained the infection to smaller areas.
- **Backups are Essential for Recovery:** Organizations with current, tested backups can typically restore their data without paying ransoms. Backups should be stored offline to prevent them from being encrypted as well.
- **End-of-Life Systems Pose Catastrophic Risks**: Many affected organizations were running outdated Windows versions that no longer received security updates.

Rootkits

Spyware operates quietly, but it still runs on top of the operating system. A rootkit is far more dangerous than spyware, though, because it buries itself deep inside the operating system's core. A rootkit is designed to escalate privileges to a root (administrator) account by exploiting system vulnerabilities. Once it gains this level of access, it modifies core system files and hides its presence from the operating system itself, making it extremely difficult to detect and remove. With root access, an attacker has complete and persistent control, allowing them to remotely execute files, spy on users, and steal data undetected.

What You'd Notice: Often nothing, and that's the problem. Rootkits are designed for stealth. You might notice some unusual system behavior, but even anti-malware software often can't detect rootkits because they operate at such a deep system level.

How to Prevent: Defense requires a layered approach. Keep systems fully patched, use advanced anti-malware with rootkit detection and behavioral monitoring capabilities, and strictly enforce the principle of least privilege to prevent malware from gaining the initial access needed to escalate. Traditional anti-malware software that only scans files often can't detect rootkits because they operate at the kernel level and hide themselves—you need advanced solutions with specialized rootkit detection capabilities and behavioral monitoring.

> **KEY POINT**: Rootkits represent one of the most serious malware threats because they give attackers complete control while remaining hidden. They're commonly used in corporate espionage, state-sponsored attacks, and advanced persistent threats.

EXAMPLE: Stuxnet (2010)

Stuxnet, one of the most sophisticated cyberattacks ever discovered, targeted Iran's nuclear program. Believed to be a joint U.S.– Israeli operation, it was a multi-stage weapon that acted as a worm, logic bomb, and rootkit. It spread through networks, used a rootkit to hide its presence on industrial control systems, and deployed a logic bomb to physically damage nuclear centrifuges.

Mobile Malware

As our lives have shifted to smartphones and tablets, so have the efforts of malware creators. Mobile malware is specifically designed to target mobile operating systems like Android and iOS. It can take any form we've discussed, from Trojans and spyware to ransomware that locks your phone until you pay. Mobile malware commonly spreads through three main vectors:

- **Malicious Apps from Unofficial App Stores**: Apps downloaded from outside the official Google Play Store or Apple App Store carry a high risk of containing hidden malware.
- **Phishing Links via SMS or Email**: "Smishing" (SMS phishing) is particularly effective, as people tend to be more trusting of links sent via text message.
- **Exploiting Vulnerabilities in Mobile Operating Systems**: Just as with desktop OSes, mobile operating systems have vulnerabilities that must be patched through regular updates.

What You'd Notice: Your phone's battery may drain unusually quickly, data usage might spike, apps could crash frequently, or you may find unfamiliar apps installed on your device.

How to Prevent: Stick to official app stores, keep your mobile OS and apps fully updated, never click on suspicious links in texts or emails, and carefully review app permissions before installation. For example, a flashlight app doesn't need access to your contacts. While iOS is generally more secure than Android, it's not immune to malware. Jailbroken iPhones are especially vulnerable, but even standard devices can be compromised by sophisticated attacks.

KEY POINT: Your phone is a primary target for attackers because it contains a treasure trove of personal data, from banking details to private messages. Apply the same security hygiene to your mobile devices as you do to your computer.

Cryptojacking

Some malware doesn't steal your data—it steals your resources. Cryptojacking malware secretly uses your device's processing power (CPU and GPU) to mine for cryptocurrencies like Bitcoin or Monero, sending the profits directly to the attacker. It can infect your computer through malicious websites that contain hidden mining scripts (known as "drive-by mining"), or through traditional phishing emails and Trojans. Unlike ransomware, cryptojacking is designed to run silently for as long as possible to maximize the attacker's profit. For example, in 2017, the popular torrent site The Pirate Bay was caught running a JavaScript-based cryptocurrency miner on its website without notifying users, causing visitors' CPU usage to spike dramatically as their computers began mining Monero without the owners' knowledge or permission.

What You'd Notice: The most common sign is a significant and persistent slowdown in your computer's performance. Your computer's fans may run constantly at high speed, and your electricity bill might increase noticeably due to the high CPU usage. Don't ignore persistent computer performance issues and assume your device is just getting old; sudden, unexplained slowness and overheating can be signs of cryptojacking malware running in the background.

How to Prevent: Use reputable ad blockers and anti-mining browser extensions, keep your anti-malware updated, and use your system's task manager or activity monitor to check for any processes consuming an unusual amount of CPU power.

> **KEY POINT**: Cryptojacking can start just by visiting a compromised website. It steals your computer's processing power and your electricity to generate money for criminals.

Artificial Intelligence (AI)-Enhanced Malware

Just as artificial intelligence (AI) is revolutionizing legitimate industries, it's also being weaponized by attackers. AI-enhanced malware uses machine learning to become "smarter," more adaptive, and better at evading detection. It can learn from its environment, mimic normal user behavior, and dynamically change its own code (polymorphic malware) to avoid being recognized by traditional signature-based anti-malware.

What You'd Notice: By design, you'd likely notice nothing until significant damage has been done. AI-enhanced malware is engineered for stealth and can adapt its behavior to appear as legitimate system activity.

How to Prevent: Defense against AI-powered threats requires AI-powered defenses. Use modern security solutions that leverage AI and machine learning for behavioral analysis, implement defense-in-depth with multiple security layers, and keep all systems up to date.

KEY POINT: Malware is no longer static. We must now assume that it can be developed by AI or embed AI to enhance its destructive capabilities, representing the next evolution in cyberthreats.

GOING DEEPER: The AI Arms Race

AI-enhanced malware has created an escalating arms race. As security companies use AI to detect threats, attackers use AI to evade those same attempts at detection. AI also lowers the barrier to entry, allowing less-skilled attackers to create more sophisticated malware than ever before. On the defensive side, AI allows security tools to analyze vast amounts of data and detect threats based on suspicious behavior, rather than just known malware signatures. This ongoing battle means the speed of attack and defense has accelerated, making proactive security measures more critical than ever.

Fileless Malware

What if malware never touched your hard drive? Fileless malware is a stealthy attack technique that operates entirely within your computer's RAM (random-access memory). It avoids leaving a traditional file-based footprint by hijacking legitimate, trusted system tools that are already on your computer, a technique known as "living off the land." Attackers commonly abuse tools like PowerShell and Windows Management Instrumentation (WMI) to execute malicious commands directly in memory. Because it uses legitimate tools and creates no new files, traditional anti-virus software that scans the hard drive is often completely blind to it.

What You'd Notice: Fileless malware is extremely difficult to spot. You might notice legitimate system processes like PowerShell.exe running unexpectedly or consuming high levels of CPU, or see unusual network traffic.

How to Prevent: Defense must go beyond file scanning. Use advanced endpoint detection and response (EDR) solutions that perform behavioral monitoring, enable robust logging of tools like PowerShell, and implement application whitelisting to control which programs are allowed to run. Modern threats like fileless malware require modern defenses focused on behavioral monitoring and real-time response, not just traditional file-scanning anti-malware software.

KEY POINT: Fileless malware evades traditional anti-malware by exploiting the trust we place in legitimate system tools. Defense requires focusing on suspicious behavior, not just suspicious files.

EXAMPLE: Astaroth Malware Campaign (2019)

In 2019, the Astaroth fileless malware campaign used phishing emails to trick users into running code that leveraged Windows' own WMI and PowerShell tools. The malware stole credentials and logged keystrokes entirely in memory, without writing a single malicious file to the hard drive, allowing it to remain undetected for long periods.

Ransomware-as-a-Service (RaaS)

The explosion of ransomware attacks in recent years is largely fueled by a criminal business model known as Ransomware-as-a-Service (RaaS). This model has dangerously democratized cybercrime, making it easy for anyone, even individuals without technical skills, to launch sophisticated ransomware campaigns.

How RaaS Works

Think of RaaS like a franchise business for cybercrime. Just as someone can open a McDonald's franchise without knowing how to develop a restaurant chain, cybercriminals can now launch ransomware attacks without knowing how to write malware. Here's how it works.

RaaS operators are skilled developers who create and maintain sophisticated ransomware platforms. They handle all the technical complexity: building the encryption software, setting up the infrastructure, creating payment portals, and even providing customer support. These operators then rent their ransomware to affiliates through underground forums and dark web marketplaces.

The affiliates are the people who actually carry out the attacks. They might lack the technical skills to create ransomware themselves, but they don't need those skills because the RaaS platform provides everything they need. Affiliates choose their targets and deploy the ransomware using methods such as phishing emails, malicious websites, or by exploiting vulnerabilities in outdated systems.

When a victim pays the cryptocurrency ransom, it's split between the affiliate and the RaaS operator. The affiliate, who executed the attack, typically keeps a majority share, while the operator receives the rest as their "commission." This lucrative profit-sharing model incentivizes both sides to participate.

Why RaaS Has Exploded

The RaaS model has been devastatingly effective because it removes the technical barrier to entry. It transforms ransomware from a specialized, skill-based attack into a scalable, commodity service. An individual no longer needs to be a master coder to be a ransomware attacker; they only need malicious intent and access to a RaaS platform. This has fueled the shift from occasional, targeted campaigns to a high-volume, global onslaught of daily attacks.

Notable RaaS Attacks

The consequences of RaaS have been staggering. Several prominent groups have successfully attacked critical infrastructure, causing widespread disruption:

- **DarkSide** attacked Colonial Pipeline in 2021, leading to fuel shortages across the U.S. East Coast and resulting in a $4.4 million ransom payment.
- **REvil** targeted JBS Foods, the world's largest meat processor, also in 2021, disrupting the food supply chain and extracting an $11 million ransom.
- **LockBit** launched an attack on the UK's Royal Mail in 2023, crippling international postal services and demanding an $80 million ransom.

These groups often operate from jurisdictions with limited law enforcement cooperation, making them difficult to dismantle. Their campaigns have collectively caused billions of dollars in damages worldwide.

Protecting Against RaaS

The defenses against RaaS-powered ransomware are the same as those against any other type of ransomware: maintain offline backups, keep systems patched and updated, train employees to recognize phishing attempts, use multi-factor authentication, implement network segmentation, and have an incident response plan ready.

> **KEY POINT**: RaaS has democratized ransomware attacks by turning them into a service that anyone can use. This explains why ransomware has gone from a specialized threat to one of the most common and costly cyberattacks today. Understanding this business model helps you recognize that ransomware threats don't just come from elite hackers, they come from an entire criminal ecosystem.

> ! **COMMON MISTAKE**: Assuming your organization is too small to be a target. RaaS affiliates use automated tools to find any vulnerable system on the Internet. In the RaaS economy, it's not about who you are, but whether you're vulnerable. Any organization with valuable data and the ability to pay is a potential target, regardless of its size.

How Hackers Deploy Malware

Having discussed common malware types, let's turn our attention to the delivery methods. Understanding how hackers deploy their malicious software is just as important as knowing what the malware does because you can use that knowledge to build defenses at the point of entry.

Phishing Emails

Phishing remains the single most common method for deploying malware. These deceptive emails are designed to trick people into either opening a malicious attachment or clicking a link that leads to a compromised website. Because this attack vector targets human psychology rather than technical flaws, it continues to be devastatingly effective. We'll take a deep dive into the various forms of phishing, including email, SMS text, and phone-based attacks, in Chapter 7.

> **KEY POINT**: Phishing remains the #1 way malware gets into organizations. No matter how strong your technical defenses are, attackers will always try to trick your users into opening the door for them.

Drive-By Downloads

A drive-by download is particularly dangerous because it requires no user interaction beyond simply visiting a compromised website. Malicious code, such as JavaScript, hidden on the site exploits vulnerabilities in your browser or its plugins to silently download and install malware in the background. This often happens in conjunction with a phishing attack, where a user is tricked into clicking a link that directs them to an infected site. Drive-by downloads are dangerous because you don't have to intentionally click or download anything—simply visiting a compromised website is enough to get infected, which is why keeping your browser and its plugins fully updated is critical.

Removable Media

Infected removable media, primarily USB drives, is a classic social engineering tactic used to breach networks. Attackers will load a USB drive with malware designed to auto-install when plugged into a computer. This method is commonly used in corporate espionage and by state-sponsored actors who may leave these drives in public areas like parking lots or lobbies, preying on human curiosity. Attackers count on curiosity to trick people into compromising their own systems, which is why you should never plug in a USB drive you find lying around. Due to this threat, most organizations have strict policies prohibiting employees from plugging unauthorized USB drives into company computers.

Trojan Horses

As we discussed earlier, a Trojan horse is a type of malware that's also a deployment method in itself. It works by disguising malicious code within a program that appears desirable, such as a free game or a system utility. The deployment happens when a user is tricked into willingly downloading and installing the software, unknowingly executing the hidden malware payload at the same time. This remains a very common and effective way for attackers to install malware onto systems.

Remote Desktop Protocol (RDP)

The Remote Desktop Protocol (RDP) is a Windows feature that allows users to connect to and control another computer over a network, which is incredibly useful for remote work and administration. However, if RDP is improperly configured and exposed directly to the Internet, it becomes a massive target. Hackers constantly scan the Internet for open RDP ports and use brute-force attacks to guess passwords, gain access, and deploy malware. RDP is a powerful tool for remote work, but it's also a favorite target for attackers. When an RDP server is exposed to the Internet without proper security controls, it's not a matter of if it'll be attacked, but when.

Bringing It All Together

Malware is one of the most persistent and dynamic threats in cybersecurity. As we've seen, it's evolved from simple viruses that required user action to autonomous worms that can cause global outbreaks, and deceptive Trojans that exploit human trust. The threat landscape now includes highly stealthy malware, like fileless variants that "live off the land" and rootkits that bury themselves in the operating system's core.

The WannaCry case study serves as a critical reminder that many devastating attacks succeed not because of brilliant hacking but because of failures in basic security hygiene, like patch management. The rise of criminal business models like RaaS has also lowered the threshold of entry for people who deploy these advanced threats, making sophisticated ransomware accessible to a wider range of attackers and dramatically increasing the volume of attacks worldwide.

Understanding the various deployment methods, from the ever-present phishing email to drive-by downloads and exploiting RDP, is just as important as understanding the malware itself. The important thing to remember is that malware threats are diverse, constantly evolving, and require layered defenses. No single security control can protect against all malware types.

Key Takeaways

- **Viruses require activation triggers** and user interaction to spread, while worms replicate autonomously by exploiting software vulnerabilities.
- **Ransomware encrypts files or entire systems** and demands payment for decryption, making offline backups the most critical defense.
- **Rootkits gain administrative access** and hide at the system level, while fileless malware operates entirely in memory using legitimate system tools to evade traditional anti-virus scans.
- **RaaS has democratized cybercrime** by turning ransomware into a franchise-like service, which explains its massive growth.
- **AI-enhanced malware represents the next evolution of threats**, using AI to adapt and evade detection, making modern, behavior-based security tools essential.
- **Phishing emails are the most common malware deployment method**, followed by other vectors like drive-by downloads, infected USB drives, and misconfigured remote access protocols like RDP.

Quick Action: Verify Your Anti-Malware Defense

In this chapter, you learned about the diverse landscape of malware threats, from viruses and worms to ransomware and spyware. The good news is that basic anti-malware hygiene can prevent the vast majority of these threats. This quick action exercise ensures your primary defense against malware is actually working, not just installed and forgotten.

Check Your Defenses

Verify Your Anti-Malware Software Is Active:

- Open your anti-malware software (Windows Security, Norton, McAfee, or whatever you've installed).
- Confirm real-time protection is enabled and running (not just installed, but actively protecting).
- Check when virus definitions were last updated (should be within 24–48 hours).

Run a Quick Scan:

- Start a quick scan. Don't run a full scan; that can take quite a long time.
- Let it run while you continue reading or doing other tasks.
- If it finds any threats, follow your software's recommendations to quarantine or remove them.

Note: If your scan finds threats, don't panic, just follow your software's recommendations to clean your system.

Your anti-malware software is your first line of defense against the threats you learned about in this chapter. Keeping it active and up to date ensures it can recognize and stop the latest malware before it harms your system.

Chapter 6. Additional Threats & Vulnerabilities

Introduction

In the realm of cyberattacks, malware accounts for only a small portion of the overall threat landscape. Beyond malware, we all face a diverse array of threats that can compromise our systems, steal our data, and disrupt our operations by targeting the fundamental building blocks of our digital world: the software and web applications we use, the networks we rely on, and the cryptographic systems we trust to keep our data safe.

In this chapter, we'll explore these additional threats and vulnerabilities. We'll begin by examining attacks that exploit flaws in code, from zero-day vulnerabilities that have no existing patch to the web-based exploits that target our applications daily. Next, we'll shift our focus to attacks against the network itself, exploring how hackers manipulate the protocols that govern our connections to intercept data or deny service. After that, we'll look at the evolving arms race between attackers and defenders, where advanced threats such as AI-driven attacks and encryption weaknesses represent the cutting edge of the cybersecurity conflict. Finally, we'll examine threats that exploit trust—insider threats and supply chain attacks. We'll examine how these attacks work, why they're effective, and what makes them particularly dangerous in today's interconnected world.

Attacks Against Code: Software's Hidden Flaws

Nearly every cybersecurity threat begins with a software flaw. Whether it's an operating system, a web application, or a mobile app, bugs and misconfigurations in code create the vulnerabilities that attackers are constantly searching for. Let's explore the most common types of code-based threats.

Zero-Day Attacks

A zero-day attack is one of the most dangerous types of cyberattacks because, by definition, no patch or hotfix exists to defend against it. The term "zero-day" refers to the situation in which software developers have had zero days to fix the vulnerability. The attack occurs on the same day the weakness is discovered, or even before the developers know the vulnerability exists. Specifically, a zero-day attack exploits a software vulnerability that the software manufacturer is unaware of. This creates a window of opportunity for attackers to exploit the flaw before the company can develop and release a patch. As you can imagine, this is an ongoing challenge in cybersecurity, as most, if not all, software is released with bugs and inherent vulnerabilities. No matter how rigorous the testing process is, it's nearly impossible to catch every potential security flaw before software reaches the market.

> **KEY POINT**: Zero-day vulnerabilities are particularly dangerous because traditional security measures such as anti-malware software and intrusion detection systems rely on known threat signatures. When a vulnerability is unknown, these defenses can't protect against it.

Bug Bounty Programs

So how do companies deal with the reality of zero-day vulnerabilities? They implement bug bounty programs. These programs create a market for ethical hackers to search for and identify security issues within software after it's been released.

These programs work by offering a bounty, either financial compensation, recognition, or a combination of both, to white-hat hackers who identify and report software flaws to the software company. The logic behind bug bounty programs is straightforward: software companies understand they won't find every single vulnerability when they release their products. They know there'll be inherent security flaws. So the hope is that white hat hackers participating in these programs will identify those issues and submit them to the company before black hat hackers can leverage them for malicious purposes.

Major technology companies such as Google, Microsoft, Apple, and Facebook all run extensive bug bounty programs, sometimes paying tens of thousands, or even hundreds of thousands of dollars for critical vulnerability discoveries. This investment in crowdsourced security testing has proven far more cost-effective than dealing with the aftermath of a successful zero-day attack.

Software Vulnerability Attacks

The reality is that when anybody writes software, whether it's an operating system, a web application, or a mobile app, it's nearly impossible to catch every bug and potential vulnerability before release. That's why software companies regularly release patches and hotfixes.

Microsoft's famous Patch Tuesday occurs on the second Tuesday of each month, when it releases security fixes and software patches for its operating system and software products. As I write this book, Microsoft released patches for 81 security vulnerabilities during their last Patch Tuesday, with two being fixes for zero-day vulnerabilities and eight classified as "Critical." This shows that even the largest and most successful software companies with virtually unlimited resources don't catch all their software bugs before release.

Another example is WordPress, the world's most popular website content management system. As of 2025, WordPress powers 43.4% of all websites on the Internet. With nearly 1.2 billion websites online, that means over 518 million sites are built on the platform. I personally run four different WordPress websites for my business. Additionally, within WordPress, to extend its functionality, I use WordPress plugins, which are software add-ons developed by third-party developers. Just like any other software, these plugins require regular updates. I monitor and update anywhere from 4 to 10 plugins on each site every week as the plugin developers release patches and hotfixes.

The main takeaway from this section is that you shouldn't assume your operating system and applications are always up to date and secure. Instead, you should proactively ensure they're kept up to date with the latest security patches. While certain operating systems and applications, like Microsoft Windows, auto-update by default, most software won't. It's your responsibility to stay on top of updates and patches. Many people delay updates because "everything's working fine," but this is a dangerous mistake. Many successful attacks exploit vulnerabilities that were patched months or even years earlier. The Equifax breach we saw in Chapter 1, for example, exploited a vulnerability that had been patched 2 months before the attack; Equifax simply hadn't applied the update.

Now, we turn to a specific type of programming flaw that's been a persistent threat for decades.

Buffer Overflows

A buffer overflow is a programming flaw that can be leveraged to introduce malicious code into a system. To understand buffer overflows, we must first understand how program memory buffers work.

Let's use a newsletter signup form on a website as an example. When a visitor enters their email address and hits submit, the web application places this information into a program memory buffer, a fixed-length block of memory allocated for the web application to temporarily store this information as it processes the request and adds it to its database. The memory buffer is designed to expect an email address of a certain length and format. If a legitimate email address is entered, everything works as intended, the data fits within the allocated memory space, the application processes it, and the email is successfully added to the database. In this "happy path," no buffer overflow occurs.

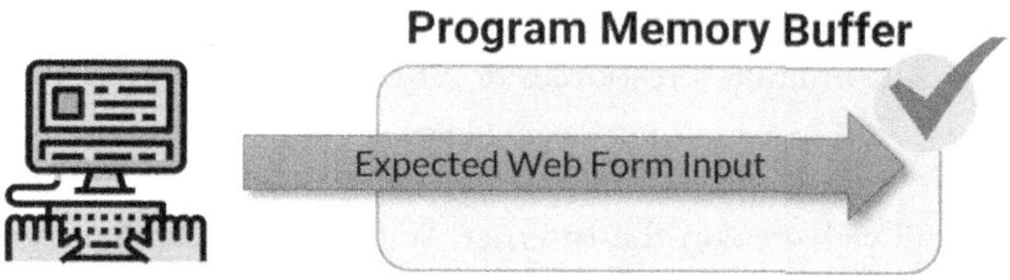

Figure 6.1: Expected Memory Buffer

However, what happens if someone types something much longer than a standard email address? If the application lacks the proper safeguards, it'll attempt to write all of the excess data into the fixed-length buffer. This is where the "overflow" happens. The data spills out of its intended container and overwrites adjacent memory locations, much like overfilling a glass of water causes it to spill onto the table. This corruption of adjacent memory can cause the application to behave erratically, crash, or create a security vulnerability that an attacker can exploit.

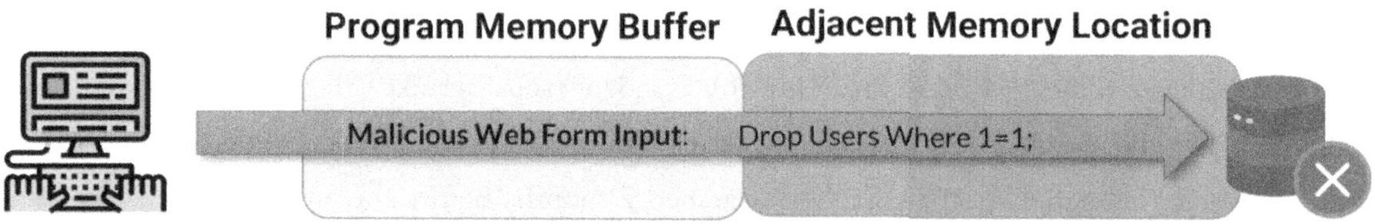

Figure 6.2: Buffer Overflow

Attackers deliberately exploit this flaw by inputting a long string of carefully crafted malicious code into web forms or other input fields. Their goal isn't just to crash the program; it's to overwrite adjacent memory with their own executable instructions. If successful, the application might be tricked into running the attacker's code, potentially allowing them to install malware or take control of the system.

Since buffer overflow attacks are commonly used against web forms and user input fields, the primary way to combat them is through data input validation. Input validation consists of safeguards that check all data before it's processed, ensuring that it's the expected type and length. If the input is too long or in the wrong format, the application simply rejects it, preventing the overflow from ever happening. It's critical to understand that client-side validation (validation performed in the user's browser) isn't sufficient on its own. Attackers can easily bypass client-side validation, which is why server-side validation is essential for true security.

While buffer overflows exploit memory management flaws, our next set of attacks exploits the logic of how web applications handle user input.

Web-Based Attacks

Now we're going to discuss web-based attacks, specifically SQL injection (SQLi) and cross-site scripting (XSS) attacks. These two attacks fall under the Open Web Application Security Project (OWASP) Top Ten list, the IT security industry's consensus ranking of the most critical web application security risks.

SQL Injection (SQLi) Attacks

For decades, the websites we interact with every day have been data-driven. From e-commerce sites and social media platforms to simple blogs, most modern web applications don't store their information in static pages. Instead, they rely on backend database servers to store and manage everything from user accounts to product inventories. My WordPress websites are a perfect example of this. When you read a blog post on one of my sites, the application isn't just pulling up a simple file; it's querying a database using a language called SQL to retrieve that specific article and display it for you.

SQL stands for Structured Query Language, and it's the programming language web developers use to communicate with databases. SQL is commonly used to retrieve, update, insert, and delete data. However, if not handled securely, SQL can also be used to delete entire databases or extract sensitive information—something a malicious user might attempt in an SQL injection (SQLi) attack. In an SQLi attack, a black hat hacker attempts to use SQL statements to their advantage by inputting malicious SQL code into a website's input field, such as a search bar or login form.

To illustrate this, we'll look at an SQL injection attack example. We'll assume a hacker has visited a website's contact page. Rather than inputting their email address in the email field, the hacker inputs malicious SQL code. For example, the hacker might input:

- **SELECT * FROM Users WHERE 1=1;**.

This SQL statement asks the database server to output everything within the Users table because the condition 1=1 will always be true.

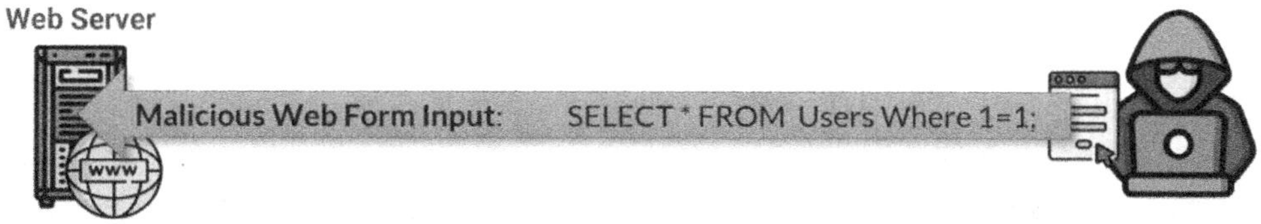

Figure 6.3: SQL Injection (SQLi) Attack

If this website and its associated database server aren't properly secured and aren't performing input validation, this SQL injection attack may be successful. All the user records within the Users database table could be displayed on the hacker's screen, potentially exposing usernames, email addresses, hashed passwords, and other sensitive information.

Another way an attacker could cause harm is by using a DELETE command, such as:

- **' OR 1=1; DELETE FROM Users WHERE 1=1; --**

This statement attempts to delete all records from the Users table. If successful, it would permanently erase all user data, including usernames, passwords, and personal information, potentially causing irreparable harm if the data isn't backed up daily.

The defense against SQL injection attacks is straightforward: Implement proper input validation and use parameterized queries (also called prepared statements) in your code. These techniques ensure that user input is always treated as data, never as executable SQL code.

Cross-Site Scripting (XSS)

Cross-site scripting (XSS) is an attack that, unlike SQL injection, targets a website's users directly rather than its database. It occurs when a hacker embeds malicious code, typically JavaScript, into either a compromised legitimate website or a malicious site they control, which then executes in an unsuspecting visitor's browser.

Think of a website like a trusted public bulletin board. An XSS attack is like an attacker sneaking a malicious flyer onto that board. When you, a trusting visitor, see the flyer, you might follow its harmful instructions without realizing the message didn't come from the board's owner. In the same way, XSS tricks your web browser into running malicious code that it believes is coming from a website you trust.

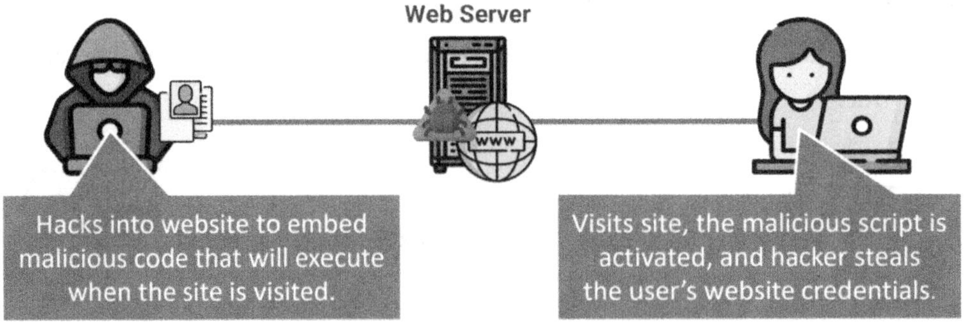

Figure 6.4: Cross-Site Scripting (XSS) Attack

Here's how it works: An attacker finds a vulnerability on a website that allows them to inject their malicious script. When a victim later visits that compromised page, their browser automatically executes the script because it trusts all code coming from that site. Once running, the script can perform a variety of malicious actions on the attacker's behalf, such as:

- Stealing the user's session cookies to hijack their logged-in accounts.
- Capturing keystrokes, including usernames and passwords.
- Redirecting the user to a phishing site.
- Performing actions on the user's behalf, like posting content or changing account settings.

Furthermore, the victim typically has no idea this is happening; the website looks and functions normally while the attack occurs silently in the background.

Case Study: Sony PlayStation Network (PSN) Breach

Background

In April 2011, Sony discovered that its PlayStation Network (PSN) had been compromised, resulting in one of the largest data breaches in history at the time. The popular gaming and media network was taken offline entirely on April 20, 2011, and remained down for 23 days, infuriating its global user base. Approximately 77 million PSN accounts and an additional 24.6 million Sony Online Entertainment accounts were compromised, exposing massive amounts of personal information and account data.

The Attack

The attack didn't rely on a single, highly advanced, or unknown exploit. Instead, the attackers exploited a chain of well-known and preventable vulnerabilities in Sony's web application infrastructure. The primary entry point was a SQL injection vulnerability. Reports at the time indicated that parts of the environment were also running outdated software, making the attackers' job easier.

Using SQL injection techniques, the hackers bypassed authentication mechanisms and gained unauthorized access to database servers containing vast amounts of user information. Once inside, they were able to access user databases and extract personal information. The stolen information included names, addresses, email addresses, birth dates, PSN login credentials, and security questions and answers.

The breach reportedly went undetected for several days, allowing attackers ample time to navigate Sony's network and extract data. Sony's security monitoring systems failed to detect the intrusion in real-time.

Impact

Sony estimated the total cost to be approximately $171 million, including identity theft protection services, legal fees, network security enhancements, and lost revenue during the outage. The company faced numerous class-action lawsuits and was fined £250,000 (around $395,000 at the time) by the UK's Information Commissioner's Office for failing to protect customer data. In total, about 77 million PSN accounts were affected, and PSN was offline for 23 days, amplifying both their financial and reputational damage.

The reputational damage was quite severe. Sony faced intense criticism for its delayed response, taking nearly a week to publicly announce the breach after discovering it. Millions of customers lost trust in Sony's ability to protect their information, and the incident significantly damaged the PlayStation brand.

Lessons Learned

- **Input Validation Prevents SQL Injection Attacks**: Proper input validation and parameterized queries may have stopped the primary attack vector that compromised Sony's systems. Organizations should validate all user input before processing it.
- **Timely Incident Notification Is Critical**: Sony's week-long delay in notifying users compounded the damage and eroded customer trust. Organizations should have clear incident response plans that include prompt, transparent communication.
- **Regular Security Assessments Identify Vulnerabilities**: The breach revealed that Sony may not have conducted adequate security testing. Regular vulnerability assessments and penetration testing may have helped identify and fix weaknesses before attackers could exploit them.

Attacks Against Networks: Exploiting How We Connect

Having discussed application flaws, let's turn our attention to attacks against the network itself. These attacks exploit the fundamental rules of networking to intercept data, impersonate devices, or deny service.

Protocol Spoofing

Protocol spoofing is the misuse of a network protocol to initiate a cyberattack against a system or network. To help you understand this concept, we'll examine two common spoofing attacks: ARP and DNS spoofing.

ARP Spoofing (or ARP Poisoning)

Before we discuss ARP spoofing, also known as ARP poisoning, we first need to understand how ARP works. ARP stands for Address Resolution Protocol, and it's designed to resolve IP addresses to MAC addresses. For example, if your laptop knows the IP address of your printer but not its MAC address, it would use ARP to resolve the MAC address. Why is this important? Remember, switches use MAC addresses for intra-LAN communication. When devices communicate within a local area network, they need to know each other's MAC addresses. So ARP plays an essential role in switch communication within a LAN.

Now that you understand how ARP works, let's discuss ARP spoofing. This is an attack that modifies the network's ARP cache to impersonate a victim computer's MAC address. Let's look at an example of a user communicating with a file server to illustrate how ARP spoofing works.

In a normal scenario, the laptop sends out an ARP request asking, "What's your MAC address?" The file server receives the ARP request and responds with its MAC address. The laptop then writes the file server's MAC address into its ARP cache, which is where the computer stores IP-to-MAC address mappings for quick reference.

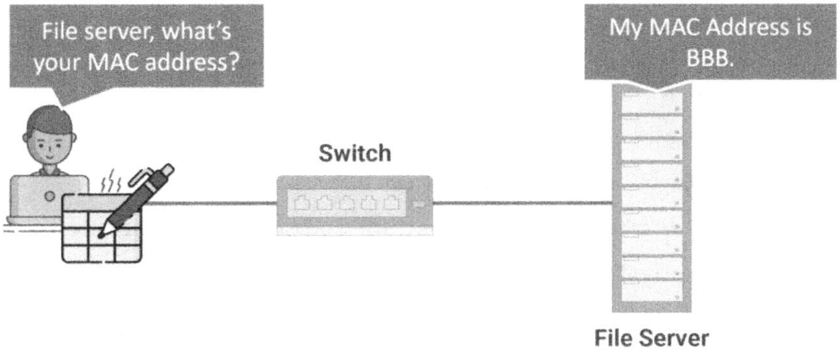

Figure 6.5: Normal ARP Request and Response

Let's now introduce a black-hat hacker into this scenario. We'll assume that the hacker has gained access to the LAN and is now attempting to impersonate the file server. ARP caches are cleared on computers every 10 to 20 minutes, so the user will eventually need to send another ARP request to the file server.

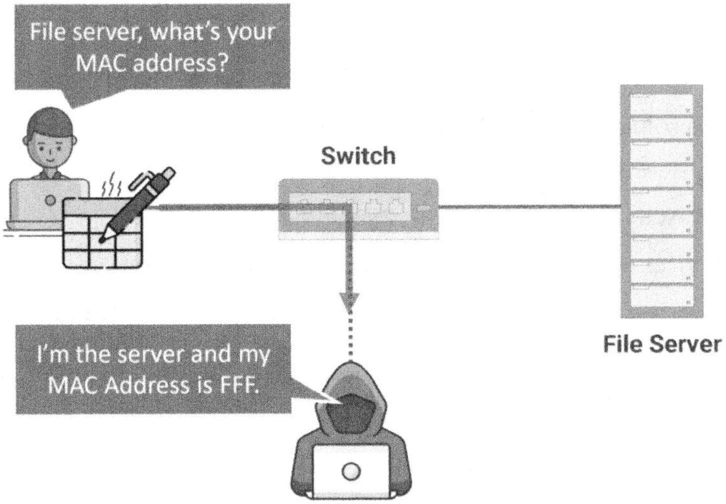

Figure 6.6: ARP Spoofing (or ARP Poisoning)

When this occurs, rather than the file server responding to the ARP request, the hacker responds instead, providing their own MAC address. The laptop will be given the hacker's MAC address and will write it into its ARP cache. From that point forward, the user will unknowingly communicate with the hacker rather than the file server.

The hacker may also perform a man-in-the-middle (MITM) attack, where they sit in the middle of the communication between the user and the file server, impersonating both MAC addresses. In an MITM attack, the hacker forwards all requests back and forth between the user and the file server while spying on and potentially modifying the data in transit. The user and the file server both believe they're communicating directly with each other, when in reality, all their traffic is flowing through the attacker. This type of attack is particularly dangerous on public Wi-Fi networks. In 2023, security researchers demonstrated how easily an attacker on a hotel's Wi-Fi network could intercept credentials from other guests—the attack took less than 5 minutes to set up using freely available tools.

DNS Spoofing

Moving on to DNS spoofing. As you already know from Chapter 3, DNS stands for Domain Name System, and it's used to translate domain names into IP addresses via DNS servers. DNS is essential because we type in domain names whenever we browse the Internet. But routers use IP addresses when routing traffic across networks, so they need to know a domain name's IP address to establish a connection. As humans, we have a tough time memorizing IP addresses. Could you imagine having to memorize the IP address of every website you visit regularly? I couldn't. So, DNS solves that problem for us, allowing us to only have to memorize easy-to-remember domain names rather than strings of numbers.

So what is DNS spoofing? It's when an attacker alters DNS records on a DNS server to redirect traffic to a fraudulent website, where further attacks can potentially occur. Let's explore how it works with an example.

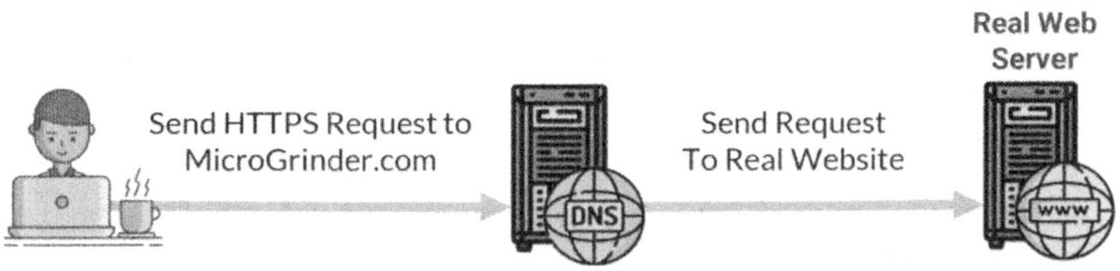

Figure 6.7: Normal DNS and HTTPS Process

As depicted above, when a person wants to visit my poker school website, MicroGrinder.com, their computer sends a request to a DNS server to get the correct IP address. The DNS server provides the real IP address, and the computer connects to the legitimate web server that hosts my website.

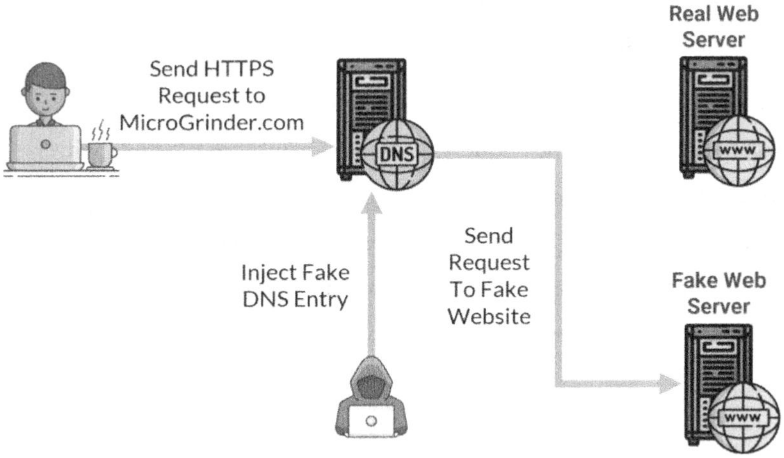

Figure 6.8: DNS Spoofing Attack

Let's now introduce a black hat hacker into the scenario. The hacker has compromised the DNS server and injected fake DNS entries that point to a fake web server designed to mimic my website, but actually created to capture user information. Now, when the person's laptop sends a DNS request for my website, it'll be given the IP address of the fake web server instead. The end user will think they're visiting my legitimate website, but they'll actually be visiting a fraudulent website where the hacker can capture login credentials, credit card information, and other sensitive data. The hacker can then launch further attacks or sell the stolen information on the dark web.

Denial of Service (DoS) & Distributed Denial of Service (DDoS) Attacks

In this section, we're going to discuss denial-of-service (DoS) and distributed denial-of-service (DDoS) attacks, two related but distinct types of attacks that aim to disrupt the availability of online services.

DoS Attacks

A DoS attack occurs when a malicious user attempts to make a server or network device unavailable by flooding it with requests.

Figure 6.9: DoS Attack

The goal is to overwhelm the server's resources, its processing power, memory, or network bandwidth, so that it can't respond to legitimate requests from actual users.

A variety of network-based attacks can be used to accomplish a DoS. One classic example is the Ping of Death, in which a hacker sends thousands of ping requests to a server until it's flooded with so many that it becomes overwhelmed and unresponsive. The ping command sends an Internet Control Message Protocol (ICMP) echo request, and the target system responds with an echo reply. As you can imagine, if a server is flooded with thousands of these requests, it'll eventually become overwhelmed and unable to process legitimate traffic. For this reason, many operating systems and firewalls now block the ICMP protocol by default to mitigate Ping of Death attacks.

DDoS Attacks

A DDoS attack is even more powerful. It's a DoS attack launched from a large number of attack machines simultaneously.

Figure 6.10: DDoS Attack

A DDoS attack may come from a coordinated group of black hat hackers or from a group of unknowingly infected computers called bots. Hackers will infect computers with malware that allows them to take remote control of the machines, making each infected computer part of the hacker's botnet, which is a network of compromised computers under the attacker's control. When the hacker is ready to launch a DDoS attack, they activate all the unsuspecting users' computers and use them collectively to carry out the attack. The victim sees traffic coming from thousands of IP addresses, making it extremely difficult to block the attack or identify the true source. Meanwhile, the owners of the compromised computers may have no idea their machines are being used to attack others.

> ! **COMMON MISTAKE**: Underestimating the business impact of an availability attack. A DDoS attack doesn't just take a website offline; it can cripple sales, damage customer trust, and serve as a distraction for a more serious data theft attack happening in the background.

Evil Twin and Rogue Access Points (APs)

Even Twin Access Points

An evil twin access point (AP) is, just like its name implies, a malicious wireless AP that advertises the same SSID (network name) as a legitimate AP to trick users into connecting to it. Once users have unsuspectingly connected to the evil twin AP and start browsing the Internet, the hacker can eavesdrop on their network traffic, intercept their credentials, and potentially inject malicious content into their web sessions.

Evil twin APs are particularly common anywhere free Wi-Fi is available, including coffee shops, airports, hotels, and conference centers. Most people will connect to the first readily available Wi-Fi network in places like these without considering the security implications. They see a network named "Airport_Free_WiFi" or "Starbucks_Guest" and assume it's legitimate, when in reality it might be an evil twin set up by an attacker sitting nearby with a laptop.

Rogue Access Points

A rogue AP is slightly different. While an evil twin is set up by an external attacker, a rogue AP is typically an unauthorized AP installed on a secure network by an insider, either maliciously or innocently. For example, an employee might install their own wireless router in their office for convenience, not realizing they've created a security vulnerability that bypasses the organization's network security controls.

For these reasons, you should always be vigilant when connecting to public Wi-Fi hotspots. Many people blindly trust public Wi-Fi and connect to networks named "Airport_Free_WiFi" without a second thought, but this is a dangerous mistake. Always verify the network name with staff and look for secure networks that require passwords. Furthermore, use a virtual private network (VPN) whenever possible, which we'll discuss in Chapter 8. A VPN encrypts all your traffic, making it unreadable to anyone intercepting it, even if you're connected to a malicious AP. The danger is real: In 2016, security researchers set up an evil twin AP at a major airport and monitored how many people connected to it. Within 2 hours, over 200 devices had connected to the fake network, and many users proceeded to check email, log into social media, and even access banking websites, all while their traffic was being monitored.

The Evolving Arms Race: Advanced Threats

This category of threats represents the ongoing battle between attackers and defenders, where new technologies are constantly being weaponized and long-trusted systems are found to have hidden flaws.

Encryption Vulnerability Attacks

Encryption is the process of scrambling readable text so that only authorized parties can view it, and it's commonly used in IT to protect our data and sensitive information. Over the years, encryption algorithms have come and gone as older ones have been cracked and rendered insecure. Hackers, both black hat and white hat, along with academic research groups, are constantly examining encryption algorithms and looking for weaknesses. Additionally, an encryption algorithm is nothing more than a mathematical function that, with enough computing power, can be cracked over time. And over time, as computers get faster—especially with the advent of quantum computing—algorithms that were once considered secure may no longer be safe.

Once an encryption algorithm is cracked, it's considered compromised and should no longer be used. A good example is the Wired Equivalent Privacy (WEP) standard, the first widespread wireless encryption standard. WEP was introduced in 1997 as part of the original IEEE 802.11 wireless networking standard. However, in 2000, its use was discouraged due to several identified security flaws. Specifically, WEP used short encryption keys and had fundamental design weaknesses that made it easy for hackers to crack. By 2004, WEP had been officially deprecated and replaced by more secure protocols such as WPA and WPA2. Today, WEP can be cracked in a matter of minutes using freely available tools—a reality we'll examine in Chapter 8's case study of the massive TJX Companies breach, where hackers exploited this exact vulnerability to steal millions of credit card numbers.

> **TECH TALK: The Wireless Encryption Arms Race**
> The transition from WEP to WPA, then WPA2, and currently WPA3, illustrates the ongoing arms race in encryption. WPA2, which uses AES encryption, remained the standard for nearly two decades before vulnerabilities like KRACK (Key Reinstallation Attack) were discovered. WPA3, introduced in 2018, addressed these weaknesses with stronger encryption and protection against offline dictionary attacks.

Artificial Intelligence (AI)-Enabled Threats

Artificial Intelligence (AI) is transforming the cybersecurity landscape in profound ways. While AI brings tremendous benefits to defenders—helping detect threats, automate responses, and analyze vast amounts of security data—attackers are also weaponizing it to create more sophisticated, scalable threats. In this section, we'll examine how AI is being used for malicious purposes and what makes these AI-enabled threats particularly concerning.

AI-Powered Social Engineering

AI, particularly generative AI, is being used to create highly convincing phishing emails, messages, and scams that adapt based on target responses (something we'll discuss in detail in Chapter 7). Attackers leverage AI to analyze personal data from social media platforms like Facebook, LinkedIn, and Twitter to craft hyper-personalized attacks that are far more convincing than traditional phishing attempts. Unlike older phishing emails, which were riddled with obvious grammatical errors and poor formatting, AI-powered phishing produces professional, compelling messages with proper grammar, an appropriate tone, and contextually relevant content. The success rates of these attacks are significantly higher because they're much harder to distinguish from legitimate communications.

> **EXAMPLE: AI-Crafted Phishing Targeting Microsoft (2023–2024)**
> In 2023–2024, Microsoft reported campaigns where attackers used highly polished phishing emails that mimicked internal and partner communications. These emails closely mirrored prior correspondence and used professional language—often produced with help from generative AI—to trick employees into entering their credentials on fake login pages. This showed how AI makes phishing more convincing and harder to spot, even for security-aware teams.

Deepfake Fraud

Deepfake technology uses AI to create realistic audio, video, and images for impersonation and fraud. This technology has advanced to the point where it can clone voices with just a few minutes of audio samples and create a convincing video of people saying or doing things they never actually said or did. Deepfakes are increasingly being used to deceive employees and executives in business settings, impersonate public figures for scams, and create fraudulent content for social engineering attacks. This technology undermines our ability to trust what we see and hear, creating new challenges for verification and authentication.

EXAMPLE: Deepfake Cryptocurrency Scam (2024)

In 2024, deepfake videos impersonating Elon Musk circulated on social media to promote cryptocurrency "opportunities." The videos closely matched his voice and mannerisms and appeared in promoted posts and re-shares, leading viewers to send money to fraudulent wallets. Losses from these Elon Musk deepfakes reached the low millions of dollars as the videos spread across platforms. This shows how deepfakes can undermine our ability to trust what we see and hear.

AI Model Manipulation

Attackers are targeting AI models themselves, either by poisoning training data, exploiting vulnerabilities in deployed models, or tricking employees into exposing sensitive data to AI systems. As organizations increasingly adopt AI tools, the security of these systems becomes a critical concern.

EXAMPLE: AI Data Exposure at Samsung (2023)

In 2023, Samsung faced a data exposure incident after employees pasted sensitive source code into ChatGPT while seeking help with coding tasks. The submissions exposed proprietary information and highlighted the risks of using generative AI tools without proper safeguards and policies. Samsung stepped in, tightened internal rules around public AI tools, and reminded employees that anything shared with external AI systems can be retained outside the company's control. The incident also prompted many organizations to implement stricter policies on what information can be shared with AI systems.

Synthetic Identities and Fake Personas

AI is also being used to create synthetic identities and fake online personas for fraud and misinformation campaigns. With AI, attackers can generate realistic profile pictures, create convincing backstories, and even automate interactions to make fake personas appear legitimate. These synthetic identities can be used to infiltrate organizations, spread misinformation, or conduct social engineering attacks at scale.

The ease with which synthetic identities can be created is alarming. Free tools and websites can generate realistic fake profile pictures, and AI can create complete fake identities with names, addresses, phone numbers, and employment histories. This makes fraud at scale more feasible than ever before and creates new challenges for organizations trying to verify the identities of people they interact with online.

EXAMPLE: AI-Made Synthetic Personas on LinkedIn (2023–2024)

In 2023–2024, attackers used AI to spin up synthetic identities on LinkedIn—complete with AI-generated headshots, polished backstories, and automated small talk that felt real. These fake "recruiters," "vendors," and "investors" connected with employees, then sent friendly follow-ups with "proposal" links or "CVs" that harvested logins or delivered malware. In multiple cases, those first clicks led to account takeovers and internal data theft. This shows how AI-made personas can slip past human judgment at scale and turn a simple connection request into a security breach.

Exploiting Trust: Insider Threats & Supply Chain Attacks

For our final category of threats in this chapter, we'll explore attacks that don't break down the door, but instead exploit the trust we place in people and partners.

Insider Threats

An insider threat is a security risk that comes from within an organization. It involves current or former employees, contractors, or business partners who have legitimate access to a company's systems and data. Think of it this way: a trusted employee has the keys to the company's secure supply room. For example, a malicious insider might use that access to intentionally steal valuable equipment. Conversely, a negligent insider might simply make a mistake, like accidentally leaving the door unlocked overnight, allowing a thief to walk in. In both cases, the company suffers a loss, but the insider's intent is what separates the two.

It's a common mistake to think that all insider threats are about malicious employees seeking revenge. In reality, accidents and carelessness are just as dangerous. Insider threats generally fall into three main categories:

- **Malicious Insiders:** Employees, contractors, or partners who intentionally steal data, sabotage systems, or leak sensitive information for personal gain, revenge, or other motives.
- **Negligent Insiders:** Employees who accidentally expose sensitive information through careless actions, such as clicking on phishing links, misconfiguring cloud storage, or losing devices containing sensitive data.
- **Third-Party Risks:** Vendors, contractors, or partners with access to company systems who may inadvertently introduce vulnerabilities or be targeted by attackers to gain access to the organization.

EXAMPLE: FedEx Cloud Storage (2018)

In 2018, security researchers discovered that a cloud storage server belonging to FedEx was left unsecured and publicly accessible by accident, exposing the personal information of over 100,000 customers. This wasn't the work of a malicious hacker but an employee who unintentionally misconfigured the cloud storage system, leaving scanned documents like passports and driver's licenses accessible. This incident highlights how a simple, careless mistake can have consequences just as severe as a deliberate attack.

KEY POINT: Insider threats are not always malicious. Unintentional negligence, like an employee falling for a phishing scam or making a configuration error, is a massive and often overlooked risk for organizations.

Supply Chain Attacks

The concept of exploiting trust extends beyond the people you employ. It also applies to the companies you do business with. A supply chain attack is a cyberattack that targets an organization by attacking its less-secure partners, vendors, or software suppliers. Imagine you run a five-star restaurant with an impeccably clean kitchen. You follow every health and safety rule. But one day, your trusted vegetable supplier delivers a batch of contaminated lettuce. Your customers get sick, and your restaurant's reputation is ruined. Even though the problem didn't start in your kitchen, a vulnerability in your supply chain led to a disaster. This is exactly how a supply chain attack works in cybersecurity.

Attackers know that large organizations often have strong defenses, so instead of attacking them directly, they target a smaller, less secure vendor in their supply chain. This could be a software provider, a contractor, or any third party that has trusted access to the main company's network or data.

In cybersecurity, supply chain risk refers to the potential vulnerabilities introduced by third-party vendors, suppliers, or service providers. These risks exist because a single compromised third party can serve as a stepping stone for attackers to gain access to the networks of all their clients. This type of attack is particularly dangerous for three reasons:

- **Expanded Attack Surface**: Every vendor you work with adds another potential entry point for attackers.
- **Lack of Direct Control**: You have limited visibility into your vendors' security practices, making it hard to know if they're truly secure.
- **Cascading Effect**: A single breach at a popular software provider can instantly affect thousands of its customers.

EXAMPLE: The SolarWinds Breach (2020)

In 2020, one of the most sophisticated supply chain attacks in history was discovered. Russian state-sponsored hackers compromised SolarWinds, a company that makes IT management software used by thousands of organizations, including government agencies and Fortune 500 companies. The hackers inserted malicious code into a legitimate software update. When SolarWinds sent the update to its customers, they unknowingly installed a backdoor that gave the attackers access to their networks.

The impact of this breach was staggering, causing widespread disruption. The primary goal was cyber-espionage, allowing the attackers to spy on sensitive internal data and communications. The victims included numerous U.S. Federal Government agencies, such as the Department of the Treasury, the Department of Commerce, and the Department of Homeland Security. In addition to government targets, thousands of private companies were also compromised. Because the backdoor was delivered via a trusted software update, the attackers operated undetected for nine months, giving them ample time to monitor internal emails and steal sensitive information. A single compromised vendor led to the breach of thousands of high-value targets worldwide.

KEY POINT: An organization is only as secure as the weakest link in its supply chain. Attackers will always look for the easiest path in, and often, that path leads through a trusted but less-secure partner.

Bringing It All Together

This chapter explored the vast threat landscape beyond malware, revealing how diverse and interconnected cybersecurity threats have become. We've seen that security requires constant vigilance at every level.

The Sony PlayStation Network breach serves as a stark reminder of what happens when multiple code-based failures—unpatched software, SQL injection vulnerabilities, and weak data protection practices—are combined, resulting in a catastrophic data breach. This highlights why fundamental security practices like input validation and regular patching remain critical, even as threats evolve.

We also saw how attackers exploit the trust inherent in our network infrastructure through protocol spoofing, DDoS attacks, and evil twin access points that turn public Wi-Fi into a hunting ground. These attacks remind us that the rules and systems that allow us to communicate can themselves be turned into weapons.

The emergence of AI-enabled threats represents a fundamental shift in the arms race between attackers and defenders, with AI being used to create hyper-personalized phishing, deepfakes, and synthetic identities that are more difficult to detect than ever before. And as we've just discussed, not all threats come from the outside. The exploitation of trust by malicious or negligent insiders and supply chain vulnerabilities, as seen in the devastating SolarWinds attack, demonstrates that trusted relationships can become powerful attack vectors.

The common thread is that security requires layered defenses and a constant awareness of these ever-evolving threats.

Key Takeaways

- **Zero-day attacks exploit previously unknown vulnerabilities** before patches are available, making them extremely dangerous. Bug bounty programs incentivize ethical hackers to find and report these flaws first.
- **Input validation is the primary defense** against code-based attacks such as buffer overflows and SQL injection; all user input must be validated and sanitized before processing.
- **Protocol spoofing attacks**, such as ARP and DNS spoofing, exploit trust in network protocols to redirect traffic, intercept communications, or impersonate legitimate systems.
- **DDoS attacks use botnets** (networks of compromised computers) to launch coordinated attacks that are significantly harder to defend against than single-source DoS attacks.
- **Regular patching is critical** because most successful attacks exploit vulnerabilities that were fixed months or years earlier; attackers rely on organizations failing to update their systems.
- **AI is being weaponized** for sophisticated attacks, including AI-powered phishing, deepfakes, and synthetic identities; these threats are more convincing, scalable, and difficult to detect than traditional attacks.
- **Insider threats exploit trusted access** from within an organization and can be either malicious or unintentional, making them difficult to detect and prevent.
- **Supply chain attacks compromise an organization** by targeting its less-secure partners and software vendors, proving that security is only as strong as the weakest link in the chain.

Quick Action: Remove Unnecessary Software

In this chapter, you learned how attackers exploit software vulnerabilities through zero-day attacks, SQL injection, and other techniques. Every piece of software on your device is a potential vulnerability, especially outdated or forgotten applications. In this quick action exercise, you'll reduce your attack surface by identifying and removing software you no longer need, making your devices more secure immediately.

Clean Up Your System

Start by making a quick list of the apps you actually use regularly. Open your device's app list on both your computer and smartphone and look through everything installed. As you review what's there, think about what you can remove.

Consider Removing:

- Apps you don't recognize (could be bundled software or potentially unwanted programs)
- Trial software you never use
- Old games or tools you haven't opened in over a year
- Multiple versions of the same type of app (do you really need 3 PDF readers?)
- Software labeled as "optimizer," "cleaner," or "booster" that you don't recognize—these are often adware or bloatware

Be thoughtful as you go. If you're unsure whether you need something, do a quick web search first to see what it does and confirm it's not a critical OS component. Start by removing obvious candidates, the things you know you don't use. When in doubt, keep it; you can always remove it later.

After uninstalling, restart your device to ensure all components are fully removed. Some programs leave background services running even after uninstallation.

Make It a Habit:

Set a reminder to review the installed apps on both your computer and phone every 6 months. Before installing new software, ask yourself: "Do I really need this?" Regular pruning keeps your attack surface small and your devices safer.

Chapter 7. Social Engineering & Email-Based Threats

Introduction

Up to this point, we've focused primarily on the technical aspects of cybersecurity: malware, network attacks, software vulnerabilities, and encryption weaknesses. But there's another dimension to cybersecurity that's equally important and, in many ways, more challenging to defend against: social engineering and the human element.

Social engineering is the art of manipulating people into divulging confidential information or performing actions that compromise security. Unlike technical attacks that exploit software vulnerabilities or network weaknesses, social engineering exploits human psychology, specifically, our natural tendencies to trust, help others, and comply with authority figures. It's the non-technical side of cybercrime, relying on deception, manipulation, and psychological tricks rather than sophisticated hacking tools.

In this chapter, we'll explore the world of social engineering and email-based threats. We'll examine how attackers use psychological manipulation to bypass even the most sophisticated technical security controls. From basic phishing emails to advanced artificial intelligence-enhanced attacks, from dumpster diving to deepfake fraud, we'll cover the most common social engineering tactics that cybercriminals employ. We'll also examine the 2020 Twitter Bitcoin Scam, a case study that demonstrates how social engineering can compromise even the most high-profile accounts and organizations.

> **KEY POINT**: Social engineering attacks succeed not because of technical sophistication, but because they exploit human psychology—our trust, helpfulness, and emotional responses. The best defense is awareness and healthy skepticism.

Social Engineering Fundamentals

Social engineering is fundamentally about exploiting human nature. While firewalls, anti-malware software, and encryption protect our systems from technical attacks, they can't protect us from our own decisions. When a single employee clicks a malicious link, shares a password over the phone, or holds a door open for an unauthorized person, they can compromise an entire organization's security.

What makes social engineering particularly dangerous is that it requires minimal technical expertise. A social engineer doesn't need to know how to write malware or exploit buffer overflows—they just need to be convincing. They need to understand human psychology and know how to manipulate people's emotions and behaviors.

All Social Engineering Attacks Share Common Elements:

- **Urgency**: "Act now or face consequences"
- **Unusual Requests**: Things outside routine procedures
- **Appeals to Authority**: "The CEO needs this immediately"
- **Requests for Information or Actions**: Things you wouldn't normally provide

If something feels off, it probably is. Trust your instincts—that uncomfortable feeling when receiving an unexpected request is often your subconscious recognizing red flags.

Conning, Deception, & Impersonation

At the heart of social engineering lies conning and deception. Social engineers are essentially con artists who use psychological manipulation to trick people into doing things they shouldn't do or revealing information they shouldn't reveal. Impersonation is one of the most common social engineering tactics. The attacker pretends to be someone they're not: an IT technician, a bank representative, a government official, or even a coworker. By assuming a trusted identity, they leverage the victim's natural inclination to trust authority figures and help others.

For example, an attacker might call an employee pretending to be from the IT help desk, claiming there's an urgent security issue that requires the employee's password to resolve. The employee, wanting to be helpful and concerned about the supposed security problem, might comply without questioning the caller's identity.

Social engineers often create a sense of urgency or put victims in a state of duress, pressuring them to act quickly without thinking critically. They might claim that an account will be closed, legal action will be taken, or a critical system will fail unless the victim takes immediate action. This urgency bypasses the victim's rational thinking and triggers an emotional response.

EXAMPLE: IRS Impersonation Scam

During tax season in the United States, cybercriminals frequently impersonate the Internal Revenue Service (IRS), calling, texting, or emailing people to defraud them financially. These scammers claim there's a problem with the victim's tax return or that they owe back taxes, threatening arrest or to take legal action unless an immediate payment is made. The IRS has stated that thousands of people have lost millions of dollars to these tax scams. The real IRS never initiates contact with taxpayers by email, text message, or social media to request personal or financial information.

Example Scenarios

Recognizing these tactics in the real world requires spotting the red flags. Here are a few common scenarios you might encounter in a work environment. Pay attention to the red flags and the recommended responses.

Scenario 1: The Fake IT Password Reset

- **The Call:** "Hi, this is Jake from IT. We're doing emergency security updates and need to reset your password. Can you tell me your current password so I can update the system?"
- **Red Flags:** This call has several warning signs. It's an unsolicited contact from someone claiming to be IT, they're asking for your current password (which IT never needs), and they're creating urgency with "emergency updates."
- **What to Do:** Hang up and call your IT department's official number to verify the request. Don't use any number the caller provides.

Scenario 2: The Manager Gift Card Request

- **The Email:** "Hi, I'm in back-to-back meetings all day and need your help urgently. Can you purchase $500 in Apple gift cards and send me the codes? I need them for client gifts. I'll reimburse you tomorrow. Thanks!"
- **Red Flags:** This email contains several suspicious elements. The request itself is unusual (gift cards for business purposes), there's urgency and time pressure, it uses an unconventional payment method, and it asks you to handle this personally rather than through normal channels.
- **What to Do:** Verify this request in person or by phone with your manager using a number you already have for them.

Scenario 3: The Executive Wire Transfer

- **The Email:** "I'm working on a confidential acquisition and need you to wire $50,000 to the following account by 3 PM today. This is time-sensitive and confidential. Don't discuss it with anyone else on the team. I'll explain everything tomorrow."
- **Red Flags:** This request bypasses the normal approval process, demands secrecy, comes as an unusual request from an executive, creates extreme urgency, and instructs you not to verify with others.
- **What to Do:** Follow your company's standard wire transfer approval process, regardless of who's requesting it. Financial procedures exist for exactly this reason.

Phishing

Phishing is a form of social engineering that uses messages, typically emails or text messages, to trick people into providing personal or sensitive information or clicking on malicious links. The term "phishing" is a play on the word "fishing," as attackers cast out bait hoping victims will bite.

Phishing attacks can take many forms, from mass emails sent to thousands of people in the hope that a few will fall for the scam to highly targeted attacks customized for specific individuals or organizations. We'll explore the full spectrum of phishing techniques later in this chapter, including spear phishing, whaling, business email compromise (BEC), and AI-enhanced phishing.

Piggybacking (or Tailgating)

Piggybacking, also called tailgating, is a physical security threat where an unauthorized person gains access to a restricted area by following closely behind an authorized person. The unauthorized person might ask an employee to hold the door or swipe them in, or they might simply follow the employee through the door without their explicit consent. This tactic exploits people's natural politeness and reluctance to confront others because most people feel uncomfortable letting a door close in someone's face, especially if that person is carrying boxes, appears to be an employee, or claims to have forgotten their access card.

For this reason, most organizations with physical security requirements implement a "no tailgating" policy, where each employee must present and swipe their own key card to enter restricted areas. Some organizations go further, implementing mantrap entrances, which are small rooms with two doors where only one door can be opened at a time, ensuring that only one person with valid credentials can enter at a time.

> KEY POINT: Politeness is a security vulnerability. It's okay to politely ask someone to use their own badge, even if it feels awkward.

Dumpster Diving

Dumpster diving is exactly what it sounds like—searching through trash and dumpsters for sensitive information that can be used for malicious purposes. While it might seem like a low-tech tactic, it can be surprisingly effective. You'd be surprised how many companies and individuals don't properly destroy sensitive documentation before throwing it away. Even when organizations have policies requiring the secure destruction of sensitive documents, some employees ignore them and simply throw them into the regular trash. When the janitor collects the trash, it all ends up in the dumpster, potentially accessible to anyone willing to search through it.

Information obtained through dumpster diving can include employee directories, internal memos, financial documents, customer lists, passwords written on sticky notes, and other sensitive information. Attackers can use this information to perform additional social engineering attacks, such as impersonation and phishing, armed with insider knowledge that makes their attacks more convincing.

Some organizations address this threat by implementing secure document destruction policies, providing shredders throughout the office, using locked bins for sensitive documents that are collected and professionally destroyed, and conducting employee training on proper document disposal.

Email Spam, Spoofing, & Basic Phishing

Email remains one of the primary vectors for social engineering attacks. Let's examine some of the fundamental email-based threats individuals and organizations face daily: spam, spoofed emails, and phishing attacks.

Spam Email

Spam email is unsolicited email, typically advertising emails, that clutters our inboxes and wastes our time. Most spam is relatively harmless—annoying but not malicious. However, spam can also be a vehicle for phishing attacks and malware distribution.

Here's one I personally received:

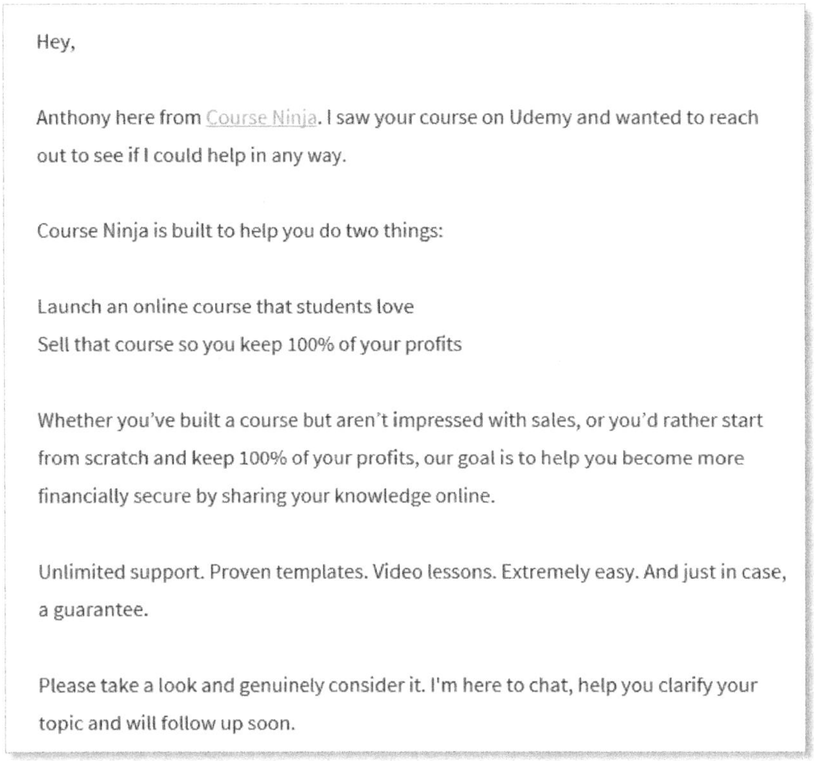

Hey,

Anthony here from Course Ninja. I saw your course on Udemy and wanted to reach out to see if I could help in any way.

Course Ninja is built to help you do two things:

Launch an online course that students love
Sell that course so you keep 100% of your profits

Whether you've built a course but aren't impressed with sales, or you'd rather start from scratch and keep 100% of your profits, our goal is to help you become more financially secure by sharing your knowledge online.

Unlimited support. Proven templates. Video lessons. Extremely easy. And just in case, a guarantee.

Please take a look and genuinely consider it. I'm here to chat, help you clarify your topic and will follow up soon.

Figure 7.1: Spam Email

Spam filters have become increasingly sophisticated at identifying and blocking spam emails, automatically routing them to spam folders. However, no spam filter is perfect, and some spam inevitably gets through to our inboxes. Additionally, some spam emails are deliberately crafted to evade spam filters by using techniques such as intentional misspellings, image-based text, and frequently changing content. The challenge with spam is distinguishing legitimate advertising from malicious phishing attempts. An email that appears to be a harmless advertisement might actually contain malware or include a link to a fraudulent website. For this reason, it's important to be cautious with all unsolicited emails, even those that appear benign.

Spoofed Emails

Email spoofing is the malicious forging of an email header to deceive the recipient into thinking the email originated from someone other than its actual source. Specifically, attackers manipulate the "From" field in the email header to make it appear as though the email came from a trusted source, such as a colleague, a company executive, a bank, or a government agency.

Email spoofing is commonly used in phishing and spam campaigns because people are more likely to open and trust an email when they believe it comes from a legitimate source. For example, you might receive an email that appears to have come from your bank, asking you to verify your account information. The email looks legitimate: it has your bank's logo, uses professional language, and appears to come from an official bank email address. However, if you examine the email header more closely, you'll discover that it actually came from a completely different email address.

Example: My Poker School Spoofing Attack

Here's a real example that happened to me. In 2019, my poker school received a spoofed email to my public email, alton@microgrinder.com, claiming that my account had been hacked. The email appeared to come from my school's own email address, as you can see in Figure 7.2, which made it seem more credible. The message demanded that I pay a $1,000 USD ransom in Bitcoin within 48 hours to prevent the supposed hacker from releasing sensitive information.

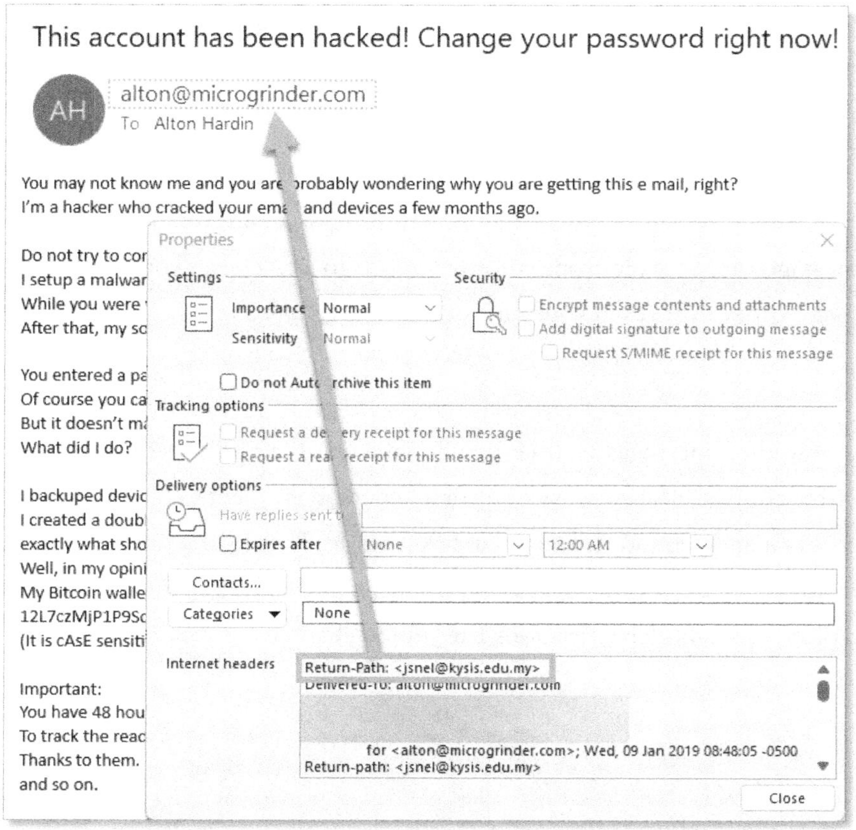

Figure 7.2: Spoofed Email Address

However, upon closer inspection of the email's header, which shows where an email actually comes from, it revealed that my email was spoofed and it actually came from jsnel@kysis.edy.my, not from my poker school's email address at all. So, my account wasn't hacked, and this was a classic example of email spoofing combined with extortion.

Most people don't know how to check email headers and will automatically assume the email came from the account listed in the "From" field. This is what makes email spoofing so effective.

What You Can Do to Protect Yourself

- **Check sender email addresses carefully:** Look at the actual email address, not just the display name.
- **Check the email header for suspicious emails:** Most email clients let you view the full header to see where the email actually originated.
- **Look for urgency tactics:** Legitimate organizations rarely demand immediate action.
- **Verify requests through alternate channels:** If someone emails asking for something unusual, call them directly to confirm using a number you already have.
- **Use multi-factor authentication:** Even if someone gets your password, they can't access your account without the second factor.

Organizations can also implement technical measures such as Sender Policy Framework (SPF), DomainKeys Identified Mail (DKIM), and Domain-based Message Authentication, Reporting & Conformance (DMARC) to help email systems verify that messages are actually from who they claim to be. These work behind the scenes to filter out spoofed emails before they ever reach your inbox.

Phishing Attacks

Phishing is the practice of sending unwanted messages to trick people into revealing personal information, such as bank account numbers, Social Security numbers, passwords, or any personally identifiable information (PII) that could enable identity theft or financial fraud. While email has historically been the primary vehicle for phishing attacks, cybercriminals now also use text messages (SMS phishing or smishing) and phone calls (voice phishing or vishing). We'll explore these modern variations later in the chapter.

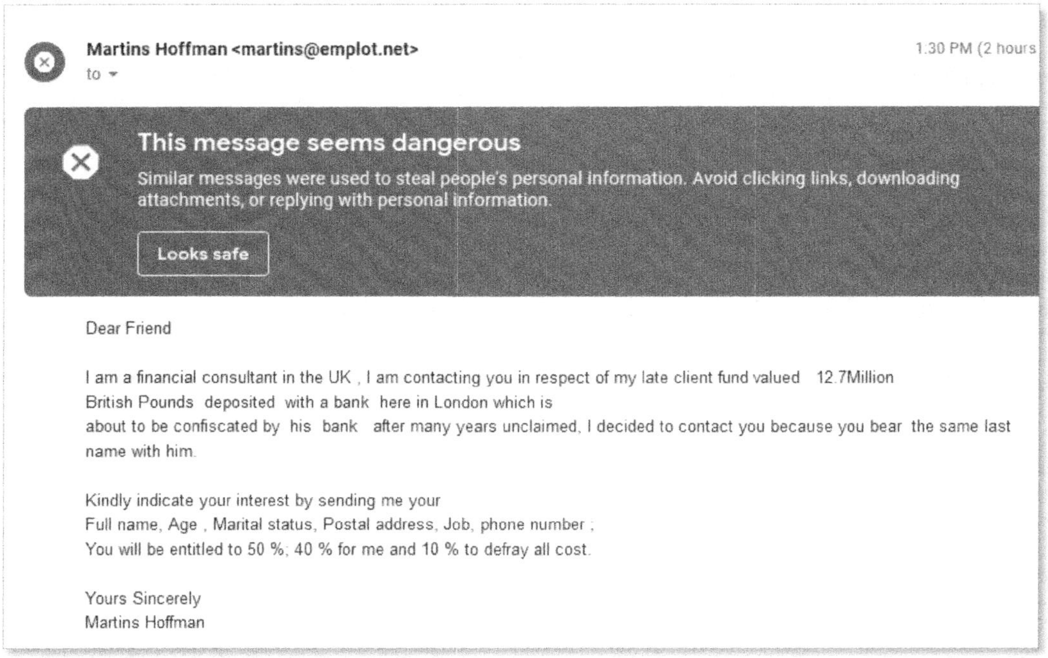

Martins Hoffman <martins@emplot.net> 1:30 PM (2 hours
to

This message seems dangerous
Similar messages were used to steal people's personal information. Avoid clicking links, downloading attachments, or replying with personal information.

Looks safe

Dear Friend

I am a financial consultant in the UK , I am contacting you in respect of my late client fund valued 12.7Million British Pounds deposited with a bank here in London which is about to be confiscated by his bank after many years unclaimed, I decided to contact you because you bear the same last name with him.

Kindly indicate your interest by sending me your
Full name, Age , Marital status, Postal address, Job, phone number ;
You will be entitled to 50 %; 40 % for me and 10 % to defray all cost.

Yours Sincerely
Martins Hoffman

Figure 7.3: Phishing Email

This is a classic phishing scam attempting to lure an unsuspecting person into providing their full name, age, marital status, mailing address, and phone number, all information that could be used for identity theft. This particular scam, where the attacker claims the victim will receive a large sum of money, has been circulating for over 15 years.

Red Flags in Phishing Emails:

- Poor grammar and punctuation
- Generic greetings like "Dear Friend" or "Dear Customer"
- Requests for personal information
- Promises of large sums of money
- Urgent threats or time-sensitive offers
- Suspicious sender email addresses
- Misspellings and awkward phrasing

Even cybersecurity professionals have been tricked by sophisticated phishing attacks. Don't assume you're "too smart to fall for this" because attackers count on overconfidence almost as much as they do carelessness. Stay vigilant with every unexpected message.

Phishing Text Message Example

Phishing text messages, better known as SMS phishing (or "smishing"), have become increasingly popular as an alternative to email. These messages often appear to be automated replies from legitimate services, making them seem more credible. We'll discuss smishing in detail later in the chapter, but here's a real example of one I received:

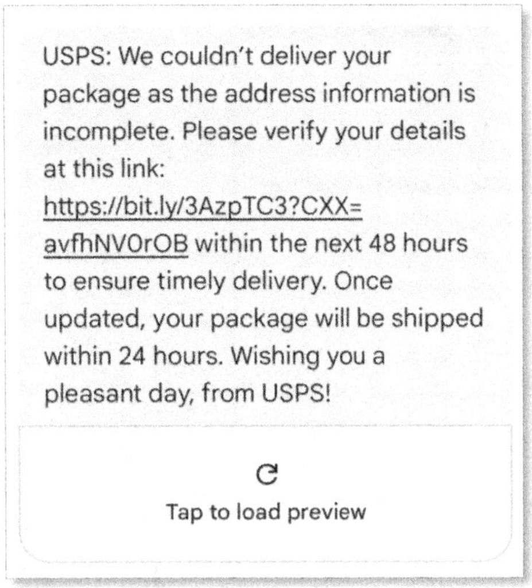

Figure 7.4: Phishing Text Message

This text pretends to be from USPS and claims my package can't be delivered until I "verify" my address. It's designed to pressure me into responding with a 48-hour deadline and includes a shortened link that would take me to a fraudulent site designed to steal my personal information or install malware.

The telltale signs here include a generic problem with no tracking number or order details, urgency meant to push a quick click ("within the next 48 hours" / "shipped within 24 hours"), a shortened URL instead of an official usps.com link, and odd phrasing and punctuation ("from USPS!") that doesn't match typical USPS notifications.

Scam Websites

Scam and fraudulent websites are designed to trick people into providing their personal information. These websites are commonly used in conjunction with phishing attacks and DNS spoofing (which we discussed in Chapter 6).

It's remarkably easy to clone a website and make a fraudulent site look exactly like the real one. The website design, logos, colors, and layout can all be copied perfectly. However, you can quickly determine if a website is genuine or fraudulent by carefully examining its URL.

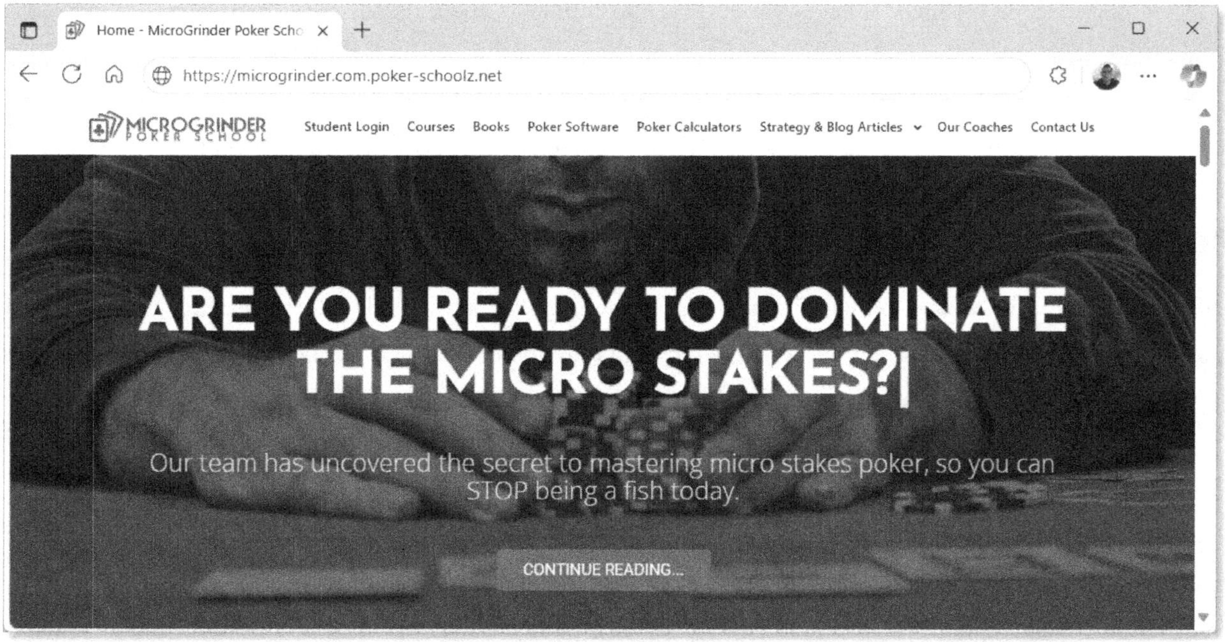

Figure 7.5: Fake URL

We can illustrate this with an example using my poker school website:

Real URL: microgrinder.com

Fake URL: microgrinder.com.poker-schoolz.net

At first glance, these URLs might look similar, but there's a critical difference. The parent domain for my real website is microgrinder.com. However, with the fake URL, the parent domain is actually poker-schoolz.net, not microgrinder.com. The "microgrinder.com" portion is just a subdomain of poker-schoolz.net.

Cybercriminals typically append URLs to make them look like the actual website. Unsuspecting victims will quickly glance at the URL in a phishing email, see " microgrinder.com," ignore the "poker-schoolz.net" portion of the address, and click the link. And when they visit the fake website, it'll look identical to the real one.

If the victim attempts to log in to the fake website, the scammer will capture their login credentials and potentially additional personal information that can be used for identity theft or account takeover.

How to Identify Fraudulent URLs:

1. Look for the primary domain, the main part of the URL, before the first single forward slash.
2. Be suspicious of URLs with multiple periods and sub-domains.
3. Check for misspellings or slight variations of legitimate domain names (e.g., "paypa1.com" instead of "paypal.com").
4. Hover over links before clicking to see the actual URL destination.
5. When in doubt, manually type the website address into your browser rather than clicking links.

Modern Phishing Techniques

While basic phishing emails remain common, cybercriminals have developed increasingly sophisticated phishing techniques that are much harder to detect and defend against. These modern phishing methods are often highly targeted, professionally crafted, and leverage advanced technologies like AI.

Spear Phishing

Unlike mass phishing campaigns that cast a wide net in hopes of catching a few victims, spear phishing is a targeted attack aimed at specific individuals or organizations. Attackers research their targets extensively, gathering personal details from social media profiles, company websites, LinkedIn, and other public sources. They use this information to craft highly personalized and convincing messages.

A spear phishing email might reference a recent project the target worked on, mention colleagues by name, or discuss specific details that only an insider would know. This level of personalization makes the attack much more convincing than generic phishing emails.

EXAMPLE: Heartland Tri-State Bank Spear Phishing Attack (2023)

In 2023, the Heartland Tri-State Bank in Kansas was shut down by regulators after it became insolvent. The bank's failure was triggered by a sophisticated spear-phishing scam, known as a "pig butchering" attack, that targeted the bank's CEO. The attackers built a relationship with the executive, eventually convincing him to wire $47.1 million to a fraudulent cryptocurrency account. The massive loss from this single, targeted social engineering attack rendered the bank insolvent.

A typical spear phishing email might look like this:

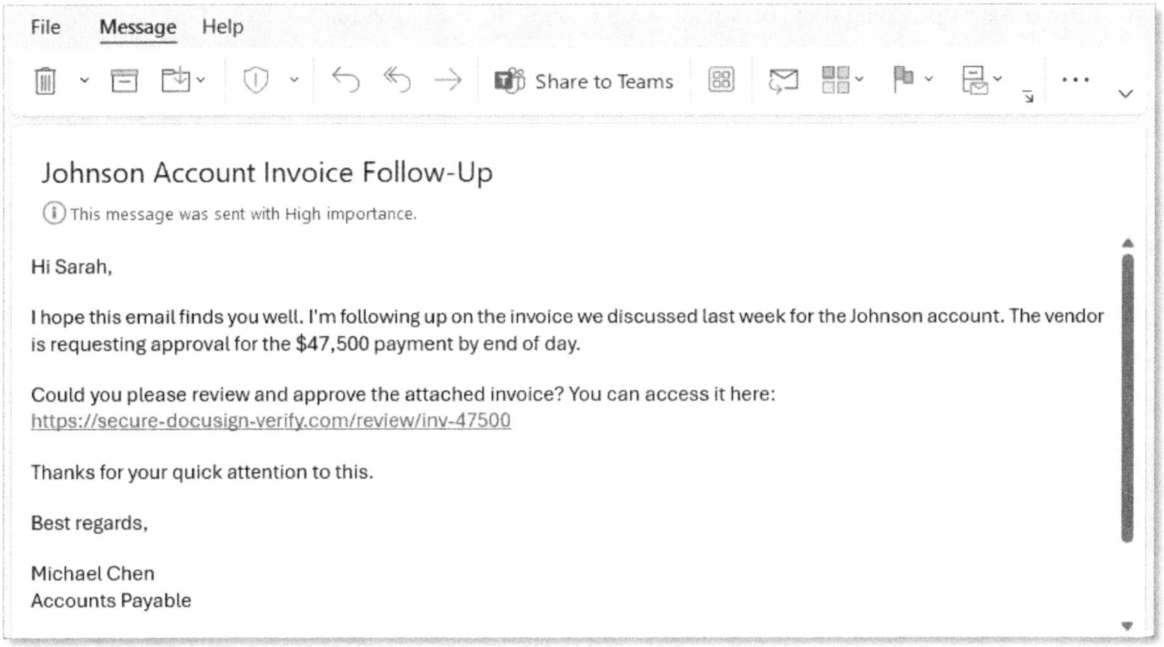

Figure 7.6: Spear Phishing Email

This email appears professional, references specific details (the Johnson account, a specific dollar amount), creates urgency (approval needed by end of day), and appears to come from a legitimate internal department. The link, however, leads to a fake login page designed to steal credentials.

Red Flags to Watch For:

- References to projects or accounts you don't recognize
- Requests that bypass normal approval processes
- Urgency that pressures you to act without thinking
- Links or attachments you weren't expecting

What to Do If Targeted:

- Verify the request through a different communication channel (call the person directly).
- Check the actual email address, not just the display name.
- Follow your organization's standard procedures, even if the request seems to come from someone you know.
- When in doubt, contact your IT or security team.

Whaling

Whaling is a form of spear phishing that specifically targets high-profile individuals–CEOs, CFOs, COOs, and other senior executives. These attacks often involve fake legal documents, financial statements, or urgent business matters that would typically require executive attention. Whaling attacks are particularly dangerous because executives often have access to sensitive information, financial systems, and permissions to authorize large transactions. A successful whaling attack can result in massive financial losses or data breaches. A whaling email might look like this:

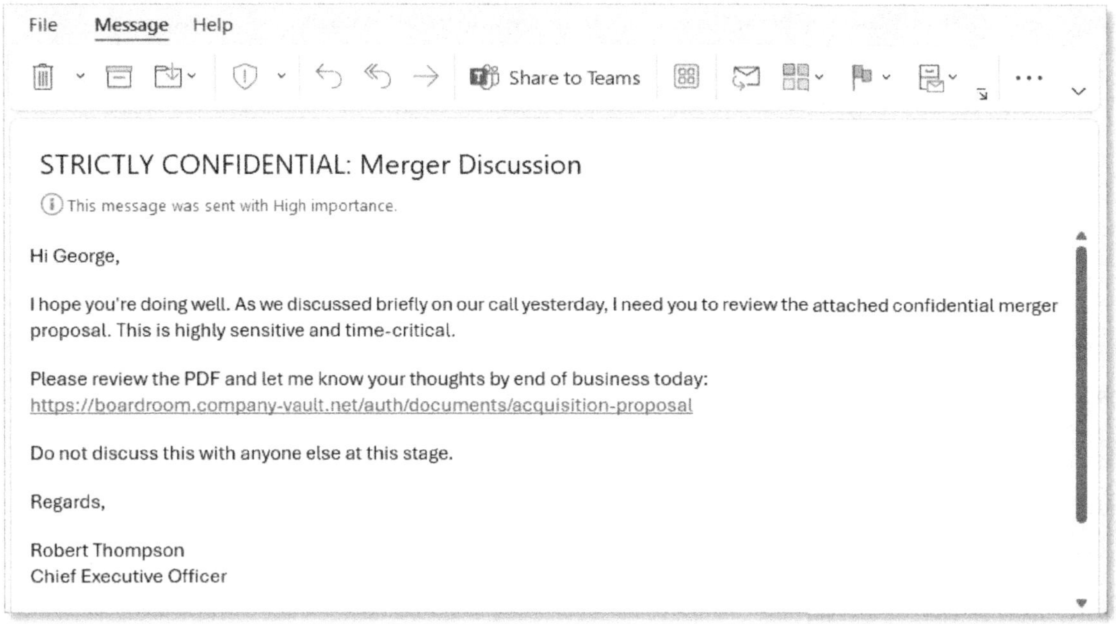

Figure 7.7: Whaling Email

This email creates a sense of importance and confidentiality, references a supposed prior conversation, includes a time constraint, and instructs the recipient not to verify with others—all tactics designed to bypass normal verification procedures.

Red Flags to Watch For:

- Unusual requests from executives, especially involving money
- Instructions to bypass standard procedures
- Demands for secrecy or confidentiality
- References to conversations that didn't happen
- Pressure to act immediately

What to Do If Targeted:

- Follow standard approval processes regardless of who's asking.
- Verify through a phone call or in-person conversation.
- Never let urgency override proper procedures.
- If instructed not to verify with others, that's a major red flag—verify anyway.

EXAMPLE: Ubiquiti Networks Whaling Attack (2015)

In 2015, attackers impersonated the CEO of Ubiquiti Networks in a whaling attack. The fraudulent communication convinced employees to execute a $46.7 million wire transfer to accounts controlled by the attackers. The attack was so convincing that multiple employees, including those in the finance department, believed they were following legitimate instructions from their CEO.

KEY POINT: Legitimate executives understand and respect security procedures. If someone claims to be an executive asks you to bypass normal processes, that's a red flag.

Business Email Compromise (BEC)

BEC is a sophisticated scam where attackers impersonate company executives or trusted business partners to trick employees into transferring money or divulging sensitive information. BEC attacks often combine email spoofing, social engineering, and extensive research about the target organization.

These attacks typically target employees in finance, accounting, or human resources departments—people who have the authority to process wire transfers or access sensitive employee information.

EXAMPLE: Toyota Boshoku Corporation BEC Attack (2019)

In 2019, Toyota Boshoku Corporation fell victim to a BEC attack that resulted in a $37 million loss. Attackers impersonated company executives and convinced employees to transfer funds to fraudulent accounts. The emails appeared legitimate, used appropriate business language, and followed normal business processes, making the fraud difficult to detect until it was too late.

Red Flags to Watch For:

- Requests to change payment account information
- Urgent wire transfer requests
- Emails that bypass normal approval workflows
- Requests for employee W-2 information or payroll data

What to Do If Targeted:

- Always verify payment changes through a known phone number (not one provided in the email).
- Follow your organization's standard approval process for all financial transactions.
- Be especially cautious with wire transfers; they're nearly impossible to reverse.
- Report suspicious requests to your security team immediately.

QR Code Phishing (Quishing)

QR code phishing, or quishing, is an emerging threat where attackers use QR codes in emails, text messages, or even physical posters to redirect victims to malicious websites. QR codes are particularly effective for phishing because:

- People can't easily see where a QR code will take them before scanning it.
- QR codes bypass many email security filters that scan URLs.
- Mobile devices, which are often used to scan QR codes, may have fewer security protections than computers.

> **EXAMPLE: Parking Meter Quishing Campaign (2023)**
> In 2023, a quishing campaign targeted users across the United States with fake QR code stickers placed on parking meters. The codes appeared to offer a convenient way to pay for parking but instead led to phishing pages designed to capture credit card details. Many drivers entered payment information on these fraudulent sites before authorities identified and removed the fake QR codes.

Quishing attacks can also occur via email, where a QR code is embedded in what appears to be a legitimate message from a bank, shipping company, or other service provider. When scanned, the code directs the victim to a fake website designed to capture login credentials or financial information.

Red Flags to Watch For:

- QR codes in unexpected emails or text messages
- QR codes on stickers that could have been placed over legitimate codes
- Requests to scan a QR code to "verify" your account or "claim" a prize

What to Do If Targeted:

- Be cautious about scanning QR codes from unknown sources.
- Check if a sticker has been placed over an original QR code.
- If a QR code takes you to a login page, manually navigate to the official website instead.

- Use a QR scanner app that shows you the URL before opening it.

Voice Phishing (Vishing)

Voice phishing, or vishing, uses phone calls to impersonate trusted organizations and trick victims into revealing sensitive information. Attackers commonly impersonate banks, tech support companies, government agencies like the Internal Revenue Service (IRS), or even law enforcement. Vishing attacks often create a sense of urgency or fear, claiming that the victim's bank account has been compromised, that they owe back taxes, or that their computer has been infected with a virus. The attacker then requests sensitive information, such as account numbers, passwords, Social Security numbers, or instructs the victim to make payments or install remote access software.

EXAMPLE: Arup Deepfake CFO Scam (2024)

In February 2024, a finance worker at the engineering firm Arup was tricked into transferring over $25 million to attackers. The attackers used AI-powered deepfake technology to create a convincing multi-person video conference call. The employee saw and heard individuals who looked and sounded exactly like the company's UK-based Chief Financial Officer (CFO) and other senior executives. Believing they were on a legitimate call, the employee followed the urgent instructions to transfer the funds.

This example demonstrates how AI technology has made vishing attacks dramatically more sophisticated and dangerous. Voice cloning technology can now create convincing audio clones with just a few minutes of authentic audio samples, which are often available from public sources such as conference presentations, podcasts, or YouTube videos.

Red Flags to Watch For:

- Unsolicited calls asking for personal information
- Pressure to act immediately
- Requests to install remote access software
- Threats of legal action or account closure
- Caller ID that matches a legitimate organization (can be spoofed)

What to Do If Targeted:

- Hang up and call the organization back using a number from their official website.
- Never provide passwords, Social Security numbers, or account numbers over the phone.
- Don't install software or grant remote access based on an unsolicited call.
- Remember: Legitimate organizations won't threaten you or demand immediate payment.

SMS Phishing (Smishing)

SMS phishing, or "smishing," uses fraudulent text messages to trick victims into clicking malicious links or providing personal information. Smishing messages often impersonate banks, delivery services, government agencies, or other trusted organizations.

Common smishing tactics include:

- Fake delivery notifications claiming a package is waiting
- Alerts about suspicious account activity requiring immediate action
- Notifications about unpaid bills or tolls
- Messages claiming the recipient has won a prize
- Fake two-factor authentication codes

EXAMPLE: IRS Tax Season Smishing Campaign (2024)

During the 2024 tax season, attackers impersonated the IRS through SMS messages, sending fake tax refund notifications to countless Americans. The messages claimed that recipients were eligible for a tax refund and needed to click a link to claim it. The link led to a fraudulent website designed to steal personal information, including Social Security numbers and bank account details.

Another common smishing scam involves fake toll payment notifications.

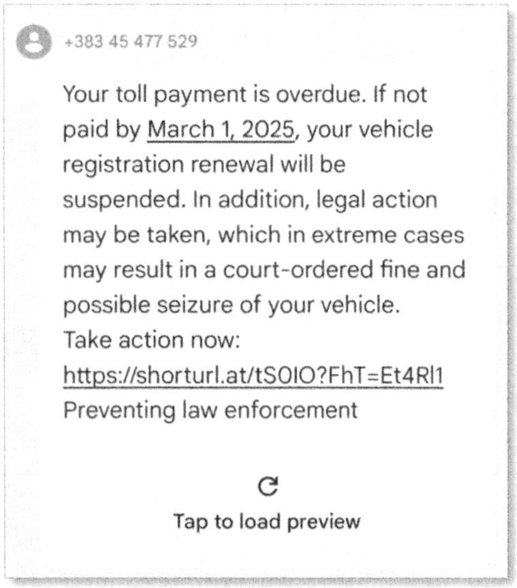

Figure 7.8: SMS Phishing Text Message

Victims receive text messages claiming they have an unpaid toll and must pay immediately to avoid legal action. The message includes a link to what appears to be an official payment website, but is actually a fraudulent site designed to steal credit card information.

Red Flags to Watch For:

- Unexpected text messages about deliveries, bills, or account problems
- Messages with shortened URLs or suspicious links
- Texts claiming you've won something you didn't enter
- Urgent threats about unpaid bills or legal action

What to Do If Targeted:

- Don't click links in unexpected text messages.
- Go directly to the official website or app instead.
- Call the organization using a number from their official website.
- Delete the message and block the sender.

Many people think text messages are more trustworthy than emails, but scammers know this and increasingly use SMS for phishing attacks. Treat unexpected text messages with the same level of skepticism you'd apply to suspicious emails.

AI-Enhanced Phishing

AI is significantly enhancing phishing attacks, making many of them more sophisticated and harder to detect. Attackers can use generative AI tools, such as ChatGPT, Claude, and other platforms, to quickly generate grammatically correct, professional phishing emails and messages that mimic the writing style of trusted individuals or organizations.

AI-enhanced phishing offers several advantages to attackers:

1. **Perfect Grammar and Spelling:** Unlike older phishing emails riddled with errors, AI-generated messages are professionally written with proper grammar, punctuation, and formatting.
2. **Personalization at Scale:** AI can analyze information about targets and generate personalized messages for thousands of victims simultaneously.
3. **Adaptive Content:** AI can generate multiple variations of phishing messages and test different approaches to determine which are most effective.

4. **Mimicking Writing Styles:** AI can analyze a person's writing style and generate messages that sound like they were written by that person.

> **EXAMPLE: Pepco Group Phishing Attack (2024)**
>
> In February 2024, the Pepco Group, a major European retail giant, lost over $15 million in a sophisticated phishing attack. The attackers used highly convincing emails that were free of errors and mirrored the tone of official correspondence. Public reporting suggested AI tools may have been used to craft these polished emails, which tricked employees into authorizing fraudulent transfers. This demonstrates how AI-enhanced scams can be far more convincing than traditional phishing emails.

The ease with which AI can generate convincing phishing content is alarming. While AI platforms have ethical safeguards designed to prevent misuse, these can often be bypassed with clever prompting. An attacker might frame their request as a "security awareness training exercise" or a "case study" to get the AI to generate phishing email templates. Here's a comparison:

Pre-AI Phishing Email:

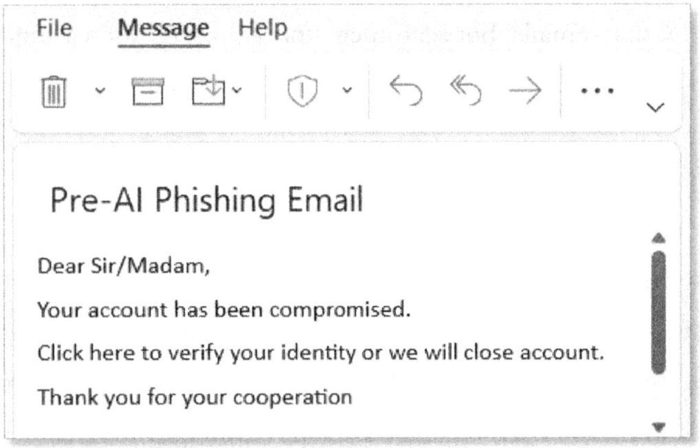

Figure 7.9: Pre-AI Phishing Email

AI-Enhanced Phishing Email:

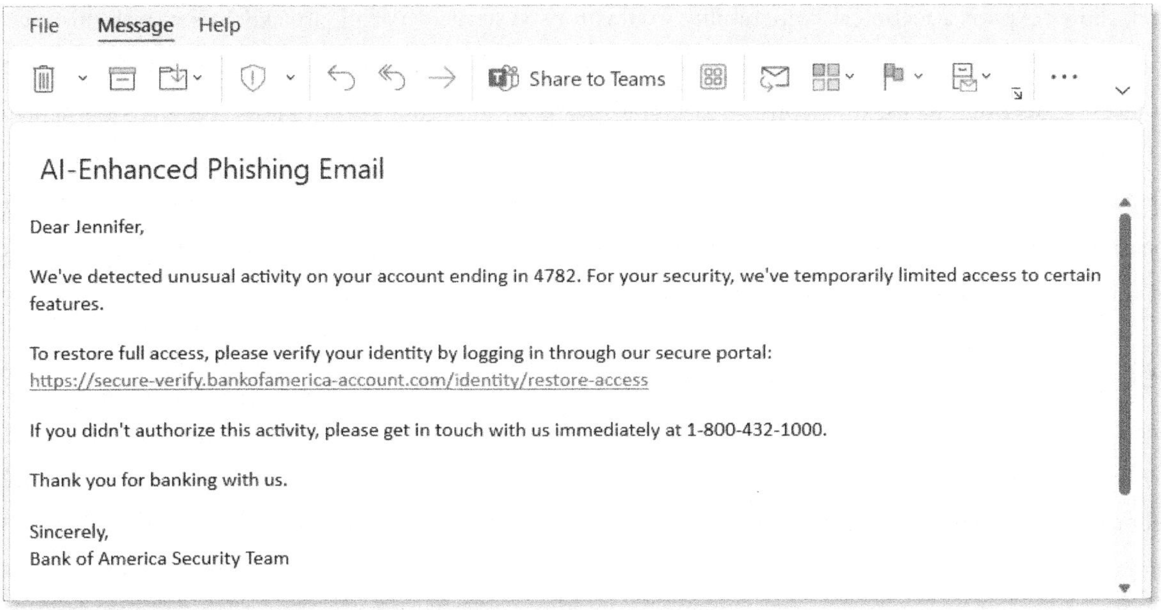

AI-Enhanced Phishing Email

Dear Jennifer,

We've detected unusual activity on your account ending in 4782. For your security, we've temporarily limited access to certain features.

To restore full access, please verify your identity by logging in through our secure portal:
https://secure-verify.bankofamerica-account.com/identity/restore-access

If you didn't authorize this activity, please get in touch with us immediately at 1-800-432-1000.

Thank you for banking with us.

Sincerely,
Bank of America Security Team

Figure 7.10: AI-Enhanced Phishing Email

The AI-enhanced version is professional, personalized, creates appropriate urgency without being overly dramatic, and includes details that make it seem legitimate. This dramatic improvement in quality means that the old advice to "look for spelling errors" is no longer sufficient. You must now verify through alternate channels, even when messages look perfect.

> **KEY POINT**: AI has eliminated the "poor grammar" red flag. Even perfectly written, professional-looking messages can be phishing attempts. Always verify unexpected requests through alternate channels.

Case Study: The 2020 Twitter Bitcoin Scam

Background

On July 15, 2020, Twitter experienced one of the most high-profile social engineering attacks in history. Attackers successfully compromised accounts belonging to some of the world's most prominent figures: Barack Obama, Elon Musk, Bill Gates, Jeff Bezos, and many others. The hackers used these hijacked accounts to post a Bitcoin scam promising to double any cryptocurrency sent to a specific wallet address. Despite the obvious nature of the scam, attackers stole over $120,000 in Bitcoin within hours.

The Attack

The attack didn't exploit a technical vulnerability in Twitter's systems. Instead, attackers used vishing (voice phishing), phone calls to Twitter employees in which they posed as members of Twitter's IT department. Using social engineering tactics, they convinced employees that an urgent security issue required immediate action. The manipulated employees provided credentials or access to Twitter's internal admin panel. Once inside, the attackers gained access to tools that allowed them to reset account passwords and disable two-factor authentication. They targeted 130 high-profile accounts, successfully taking control of 45 of them, and posted identical Bitcoin scam messages from the compromised accounts.

Impact

While the direct theft was "only" $120,000 in Bitcoin, the potential for damage was far greater. If attackers had chosen to manipulate stock markets, spread misinformation, or cause geopolitical incidents, the consequences could have been catastrophic. Twitter's credibility as a secure platform was severely damaged. The incident raised serious questions about the security of social media platforms and their internal access controls. Three individuals were arrested and subsequently convicted: a 17-year-old from Florida who was sentenced to three years in prison, and two accomplices from the UK and Florida.

Lessons Learned

- **Employees Are Vulnerable to Social Engineering**: Even sophisticated companies with advanced security systems can be compromised when attackers target employees through manipulation and deception rather than technical exploits.
- **Privileged Access Requires Strict Controls**: Organizations should implement strong safeguards around internal admin tools, including multi-factor authentication, access logging, and verification procedures for sensitive actions.
- **Security Awareness Training Is Essential**: Regular training helps employees recognize social engineering tactics such as vishing and understand proper verification procedures before granting access or issuing credentials.
- **Human Vulnerabilities Can't Be Patched**: Unlike software vulnerabilities, human susceptibility to manipulation requires ongoing education, clear policies, and a security-conscious culture throughout an organization.

KEY POINT: The Twitter hack succeeded not because of technical sophistication, but because attackers understood human psychology and exploited employees' desire to be helpful during a supposed emergency.

Bringing It All Together

This chapter shifted our focus from technical vulnerabilities to the human element, the most challenging aspect of cybersecurity to defend against. Social engineering succeeds because it exploits our natural tendencies to trust, help, and respond to urgency. These psychological vulnerabilities can't be "patched" like software; they require constant awareness and training.

The evolution of phishing from simple, error-filled emails to sophisticated, AI-generated spear phishing, whaling, and BEC attacks shows how attackers are constantly refining their methods. Modern attacks have also expanded beyond email to include vishing (voice), smishing (SMS), and quishing (QR codes), often using deepfakes and AI-cloned voices to appear legitimate.

The 2020 Twitter Bitcoin Scam serves as a critical lesson: even the most secure organizations can be compromised when attackers manipulate employees. The common thread is that technology alone isn't enough. The most critical defenses are a healthy sense of skepticism, a habit of verifying unusual requests through a separate channel, and a culture that encourages employees to prioritize security.

Key Takeaways

- **Social engineering exploits human psychology**, trust, helpfulness, and fear, making awareness and training your most critical defenses.
- **AI has dramatically increased phishing sophistication**, generating convincing, personalized messages that no longer have obvious errors like poor grammar.
- **Modern phishing is highly targeted**, with spear phishing aimed at individuals, whaling aimed at executives, and BEC aimed at employees with financial authority.
- **Attack vectors now extend beyond email** to include vishing (voice), smishing (SMS), and quishing (QR code) attacks.
- **Email spoofing forges sender addresses** to impersonate trusted sources; always verify unexpected or unusual requests through a separate communication channel.

- **The 2020 Twitter Bitcoin Scam** was a high-profile success for social engineers, who used vishing to trick employees into providing internal system access.

Quick Action: Train Your Spam Filter

This chapter showed how attackers use social engineering and phishing emails to trick you into giving away sensitive information or installing malware. No filter is perfect, but your email's spam filter can become a strong first line of defense if you help it learn what to catch. Marking suspicious emails as spam teaches your email system to recognize and block similar threats before they reach your inbox. In this quick action exercise, review your personal email inbox and train your spam filter to catch the phishing signs from this chapter.

What to Do

Check your personal email inbox for emails that exhibit the warning signs we discussed in this chapter. These often include:

- **Urgent Requests for Personal Information:** Messages that pressure you to act immediately or claim your account will be closed
- **Unexpected Attachments or Links:** Files or links you weren't expecting, especially from unknown senders
- **Emails from Companies You've Never Done Business With:** Messages claiming to be from services you don't use
- **Messages with Poor Grammar or Odd Formatting:** Awkward phrasing, spelling errors, or unprofessional appearance

When you spot a suspicious email, don't just delete it. Mark it as spam or junk using your email service's spam reporting feature. This is usually a clearly labeled button or option within your email client.

Do this consistently for a week, and you'll notice fewer suspicious emails making it to your inbox. Your spam filter learns from your actions and gets smarter over time.

PART 3 CONCLUSION

Bringing It All Together

You've now explored the diverse and constantly evolving threat landscape that security professionals defend against. You've learned how malware, from viruses and worms to ransomware and rootkits, works to compromise systems and steal data. You've discovered how attackers exploit vulnerabilities in software, manipulate network protocols, and launch sophisticated web-based attacks. Most importantly, you've seen how social engineering bypasses even the strongest technical defenses by taking advantage of human psychology through phishing, pretexting, and manipulation.

The case studies throughout these chapters—WannaCry (Chapter 5), Sony (Chapter 6), and Twitter (Chapter 7)—aren't included to discourage you. They were chosen to show you exactly why this work matters and to demonstrate a critical truth: Most breaches succeed not because of sophisticated zero-day exploits, but because of failures in basic, fundamental security practices. The WannaCry ransomware attack occurred because many organizations failed to install a security patch that Microsoft had released two months earlier. The Sony breach exploited SQL injection and well-known and preventable vulnerabilities in Sony's web application infrastructure. The Twitter hack used simple voice phishing to trick employees. These weren't unstoppable attacks; they were preventable failures in fundamental security practices.

Now it's time to shift from understanding threats to implementing the defenses that stop them. Part 4 will show you the specific tools, technologies, and practices that transform security knowledge into security action.

PART 4: IMPLEMENTING DEFENSES

Overview

Understanding threats is essential, but stopping them requires action. This section bridges the gap between security knowledge and security practice, showing you the specific tools, technologies, processes, and organizational structures that protect modern organizations. We'll start with fundamental countermeasures in Chapter 8, the essential security controls that protect your identity, devices, and network communications. Chapter 9 shifts to security operations, exploring how organizations detect threats in real-time, respond to security incidents, and proactively test their defenses. Chapter 10 examines the management and cultural aspects of security, showing you how organizations create the policies, processes, and training programs that make security sustainable. Finally, Chapter 11 addresses three critical modern challenges: building security into applications from the beginning (DevSecOps), ensuring organizations can continue operating when attacks succeed (Business Continuity), and protecting the diverse technologies employees use today—from mobile devices and IoT to cloud infrastructure. This section transforms you from someone who understands security to someone who knows how to implement it:

- Technical controls that protect systems and data
- Operational processes that detect and respond to threats
- Management frameworks that ensure security supports business goals
- Strategies for building security into applications and maintaining operations during disruptions

By the end of Part 4, you'll understand not just what security controls exist, but how they all work together as a single, integrated defense system.

Chapter 8. Fundamental Threat Countermeasures

Introduction

In the last few chapters, we explored the vast landscape of threats, from malware and network attacks to social engineering. Now, we'll shift our focus from how attackers break in to how we keep them out.

This chapter is your introduction to the essential tools and strategies you'll use to build a strong defense. We'll be building on the core principles from Chapter 4, showing you how each countermeasure acts as another layer in your security armor. We'll cover the foundational controls that protect your identity, such as multi-factor authentication and password best practices, before moving to the defenses that secure your devices and network. These include endpoint protection, encryption, firewalls, intrusion detection and prevention systems, DMZs, honeypots, VPNs, and wireless network security. These are all the various means by which we can build up a defense in depth strategy to protect our digital assets.

Multi-Factor Authentication

Multi-factor authentication (MFA) is a layered approach to verifying your identity. It requires you to prove who you are in more than one way before gaining access to a system.

Historically, when we logged into a system or website, we typically just needed our username and password. That was single-factor authentication. But as hackers have gotten more sophisticated, so have software developers and cybersecurity experts. Over the past several years, many applications and online sites now require multi-factor authentication, which is a combination of two or more of the following factors:

- **Something You Know**: Password, Passphrase, or PIN
- **Something You Have**: Smart ID card, ATM card, or Smartphone for One-Time Codes
- **Something You Are**: Biometrics like Fingerprints, Palm Scans, or Eye Scans

Figure 8.1: Multi-Factor Authentication

The whole goal of MFA is to combat malicious users who are trying to steal usernames and passwords to log into systems. Think of it like accessing a safe deposit box at a bank. You need your physical key (something you have), but that alone isn't enough. You also have to present your ID to the banker, who verifies your identity (something you are). Only with both the correct key and verified proof of identity will you be granted access. Even if a thief steals your key, they can't get into your box because they can't prove they're you.

> **KEY POINT**: Multifactor authentication (MFA) dramatically reduces the effectiveness of credential theft. Even if attackers steal your password through phishing or a data breach, they still can't access your accounts without the additional authentication factors.

Examples of Multi-Factor Authentication

The beauty of multi-factor authentication is that you're probably already using it in your daily life without even realizing it:

- **Corporate Login**: When you log into your work laptop, you might insert your company-issued smart card (something you have) and then enter your password (something you know).
- **ATM Withdrawal**: When you withdraw money from an ATM, you must insert your debit card (something you have) and enter your PIN (something you know).

What you'll typically find is that most organizations and online services now require at least two forms of authentication; some even require three. The entire premise of MFA is to make it harder for hackers to bypass your authentication measures. However, not all MFA factors are created equal. While SMS text messages are far better than no MFA at all, they're the most vulnerable option because they can be intercepted through SIM-swapping attacks. Unfortunately, many services only offer SMS, leaving you with no other option. But when you have a choice, always select authenticator apps, such as Google or Microsoft Authenticator, or hardware security keys, such as YubiKey, as they provide much stronger protection because they're not vulnerable to SIM swapping.

Password Best Practices

After MFA, a strong, unique password is your next line of defense. Passwords should be strong, but understanding "why" helps you create better ones. When we create a new password for an online service or application, we're typically given a minimum character count (modern standards recommend at least 12 characters) and told to use a combination of uppercase and lowercase letters, numbers, and special characters. This all comes down to how hackers crack passwords.

Understanding Brute Force Attacks

The most fundamental way a hacker can crack a password is with a brute force attack, which tries every single combination of a password until it guesses the right one. Your goal, therefore, is to make the pool of possible combinations so enormous that it would be practically impossible for a computer to guess the correct one.

Here's how both complexity (using different character types) and length dramatically increase a password's strength. We'll compare three types of passwords:

- **A Short, Simple Password**: An 8-character password using only uppercase and lowercase letters (e.g., PasSworD) has about **53 trillion** possible combinations. While that sounds like a lot, modern computers can crack it relatively quickly.
- **A Longer, Complex Password**: Now, let's follow modern standards by increasing the length to 12 characters and adding numbers and symbols (e.g., P@ssw0rd!234). The number of possible combinations skyrockets to over **400 sextillion**. This is a massive leap in security that makes the password exponentially harder to crack.
- **A Long Passphrase**: Finally, let's look at a long but simple passphrase like ilovepepperonipizza (I really do love pepperoni pizza). This 19-character password uses only lowercase letters, but it has over **1.6 octillion** possible combinations.

So what does this tell us? Moving from a short password to a longer, more complex one provides a significant improvement in security. Taking the next step to an even longer passphrase makes your password virtually unbreakable against brute-force attacks, while also being easier to remember. This is exactly why security experts and organizations are increasingly recommending the use of passphrases, which we'll discuss in the next section. But here's the challenge: even knowing all this, how are you supposed to manage dozens of these strong, unique passwords? That's where modern password management comes in.

Modern Password Guidelines

Now that you understand what makes a password strong, you might be wondering, "How am I supposed to remember dozens of unique, long, and complex passwords for every website I use?" The answer is simple: you don't. In this section, we'll explore the two modern solutions that solve this problem: password managers and passphrases.

Use Password Managers

One of the most effective solutions for managing dozens of unique and complex passwords is to use a password manager. A password manager is a software application designed to securely store and manage all of your login credentials in an encrypted digital vault that only you can access. Not even the company that provides the service can open it. You only need to remember one strong master password to unlock this vault, giving you access to all of your other passwords.

Password managers come in two main flavors: online (cloud-based) and offline (local). Online solutions, which are the most common, store your encrypted vault on the provider's servers. This allows you to access your passwords from multiple devices; for example, a password you save on your laptop will automatically sync and be available on your iPhone. Offline solutions, on the other hand, store your vault as a file directly on the device where the password manager is installed. This vault isn't automatically shared with your other devices, which some users prefer for maximum control over their data.

Once set up, password managers make your digital life both easier and dramatically more secure. Their main benefits include:

- **Generating Strong Passwords:** They can automatically create long, random, and complex passwords for every new account, so you don't have to create and remember them yourself. Instead of using Summer2024! for multiple accounts (a common but dangerous practice), a password manager generates and remembers unique passwords, such as &e@&CF7@PhgDfK2KL8ah, for each account. If one website gets breached, your other accounts remain secure because each uses a different password.
- **Secure Storage:** All your credentials are stored in a heavily encrypted vault, protected by your master password.

- **Auto-Filling:** Most password managers include browser extensions and mobile apps that automatically and securely fill in your username and password on websites and apps. This isn't only convenient but also helps protect you from phishing sites, as the auto-fill won't work on a fraudulent website with a different URL.

Popular options include 1Password, Bitwarden, and LastPass, with some password managers offering free versions for personal use.

Consider Passphrases

As we saw in the last section, a long passphrase like ilovepepperonipizza can be astronomically stronger than a short, complex password, and it's much easier for humans to remember. For passwords you need to remember, such as your password manager's master password, consider using a passphrase—a sequence of random words strung together.

For example, the passphrase correct-horse-battery-staple is much easier to remember than Tr0ub4dor&3 but actually provides better security due to its sheer length. The key is making passphrases long (at least 15–20 characters) and, most importantly, random.

> ! **COMMON MISTAKE**: Creating passphrases from predictable phrases like song lyrics, movie quotes, or common sayings. Attackers use dictionaries of common phrases in their cracking attempts. You should use truly random words instead.

General Best Practices

Don't Write Down Passwords

You'd be surprised how many people, even in professional environments, still rely on a sticky note attached to their monitor or a small notebook in their desk drawer to manage their passwords (my parents are culprits, even though I tell them not to). This is often easier said than done, especially in business environments where people have access to multiple systems with different credentials. But a written password completely bypasses any digital security you have in place. The solution is to use single sign-on or a password management solution. The same goes for your personal accounts: use a password manager rather than writing down your passwords.

Don't Share Passwords

This is primarily applicable in business environments. In a work setting, you should never share your password with anyone, including your manager or IT staff. This is because your user account is tied directly to you, and everything done with it is logged. If someone does something wrong, incorrect, or malicious using your shared credentials, it comes back to you.

Of course, your personal life is a different story. Sharing a password for a joint online account, such as Netflix or your electricity bill, with a spouse or partner is a normal part of managing a household. The key is to apply the "no sharing" rule strictly to your individual and professional accounts.

When to Change Your Passwords

For years, the standard advice was to change your passwords every 90 days. You'll still find many organizations that enforce this policy for compliance reasons. However, modern security guidance has shown that this practice can actually make us less secure.

While it might sound counterintuitive, forcing frequent password changes often leads to "password fatigue." Instead of creating a new, strong password, people tend to make small, predictable changes, like changing Summer2024! to Fall2024! or writing their passwords down to keep track of them. These habits ultimately make accounts easier for attackers to compromise.

The modern best practice is to create a very long, unique password or passphrase for each account and change it only when you have reason to believe it's been compromised. You should change your password immediately if:

- The service you use announces a data breach.
- You see suspicious activity on your account.
- You learn that your password was exposed in another company's data breach (and you reused that password).

> **GOING DEEPER:** This shift in thinking is championed by organizations like the National Institute of Standards and Technology (NIST). Their guidance suggests that focusing on creating long, strong, and unique passwords from the start provides better security than forcing constant, small and predictable updates.

Endpoint Protection

Endpoint protection is another name for system hardening, which is the process of reducing a system's attack surface. When we talk about endpoints, laptops, desktops, smartphones, tablets, servers, and so forth, the attack surface refers to any means by which a hacker can potentially get into a system and compromise it.

The goal of system hardening is to reduce the attack surface, making it smaller. Every unnecessary service, open port, or user account represents a potential entry point for attackers, and system hardening systematically eliminates these entry points. Think of it this way: It's like locking all the unused doors and windows of a house so that a burglar has fewer ways to break in. The smaller the attack surface, the less vulnerable that system will be to potential cyberattacks.

Ways to Harden Systems

System hardening might sound like a complex technical task, but it's really about a series of straightforward, common-sense steps. Each step is designed to close a potential door that an attacker might try to open. Here are the most common and effective ways to harden a system.

Figure 8.2: System Hardening

Disable Unused Physical Ports

This refers to physical ports on a device, such as USB ports. Disabling these ports is a direct countermeasure against certain social engineering attacks we discussed in Chapter 7. For instance, an attacker might leave an infected USB drive in a company parking lot, hoping a curious employee will pick it up and plug it into their computer. For physical ports, you can disable them in the system's settings or unplug them from the motherboard to prevent them from working. This prevents a curious employee from plugging in a found thumb drive and unknowingly installing malware.

Disable Unnecessary Services & Protocols

Every program and service running on a computer is a potential target for attackers. Even if you aren't actively using a service, a vulnerability in its code could give an attacker a foothold.

To communicate over a network, these services must "listen" on a specific, numbered logical port. Think of a logical port as the unique phone extension for a department in a large company. The web services department might be at extension 443, while the remote access department is at extension 3389. If your company doesn't have a remote access department, you wouldn't want that phone line to be active.

A web server, for instance, only needs to run web server software, not file-sharing services, print services, or remote desktop services. Disabling everything except what's absolutely necessary dramatically reduces the attack surface. For this reason, the most effective way to secure these ports is to disable the underlying services you don't need. When you shut down a service, the port it was listening on effectively closes, removing that potential entry point for an attacker.

> **TECH TALK:** Logical ports are numbered communication channels that services use to send and receive network traffic (ranging from 0-65535). Common examples include port 80 for HTTP, port 443 for HTTPS, and port 3389 for Remote Desktop Protocol. By disabling the service, you block the corresponding port and prevent attackers from exploiting it.

Disable or Delete Unnecessary User Accounts

This is especially important in business environments and relates to account management. When employees leave the organization, their accounts should be disabled or deleted immediately. It's also a best practice to regularly review systems for any guest or default accounts that are no longer in use and disable or delete them. These are often called "orphaned accounts," and they provide a convenient target for attackers.

The impulse is often to simply disable an old account instead of deleting it. While disabling is a good first step, leaving these accounts on a system indefinitely creates a risk, as attackers can sometimes find ways to re-enable them. The best practice is to have a clear policy: if an account is no longer needed, it should be deleted. However, some legal and regulatory standards require organizations to maintain audit records. In these cases, accounts should be disabled, stripped of all access permissions, and monitored closely to ensure they remain inactive.

Keep Systems Up-to-Date

As we saw with massive breaches like Equifax (Chapter 1) and WannaCry (Chapter 5), a single unpatched system can be all an attacker needs to cause a catastrophic compromise. Keep systems up to date with the latest security patches. This relates to patch management and is critical for protecting against known vulnerabilities.

Use Anti-Malware Software

Anti-malware software is your system's immune system, designed to detect and remove malicious code. Modern operating systems like Windows and macOS come with pre-packaged anti-malware software. However, if you want more robust features and granular control, you can purchase additional anti-malware software from third-party vendors.

Regardless of which software you use, the most important thing is to ensure it's always running and up to date. Anti-malware programs rely on a constantly updated database of "signatures"—think of them as digital fingerprints for malware—to identify known threats. Without regular updates, your software won't be able to protect you from the latest attacks.

> **GOING DEEPER**: Modern anti-malware software uses multiple detection methods to stay ahead of attackers. Signature-based detection works by scanning files for patterns, or "signatures," that match a database of known malware. Heuristic or behavior-based analysis, on the other hand, monitors programs for suspicious actions, such as an application trying to modify critical system files or encrypt your data. This allows modern tools to detect new, previously unknown malware.

Employ a Software Firewall

A software firewall, also called a host-based firewall, runs directly on your computer and acts as a personal security guard for that specific device. It monitors all network traffic coming into and going out of your computer, blocking potentially malicious connections based on a set of security rules. You should ensure this feature is always enabled on your operating system. We'll discuss network firewalls in detail later in this chapter, but a host-based firewall provides a crucial second layer of defense, protecting your device from threats that might already be inside your local network.

Enable Disk Encryption

Disk encryption is your safety net in case a device is physically lost or stolen, or if it's hacked. It uses a process called encryption to scramble all the data on your hard drive, transforming it into unreadable gibberish that can only be unlocked with your password. We'll discuss encryption in detail in the next section of this chapter, but for now, the key takeaway is that you should consider enabling full-disk encryption using the tools built into your operating system, such as BitLocker for Windows or FileVault for Mac. While powerful, be aware that if you lose your password and your recovery key, you could be permanently locked out of your data.

Use Complex Passwords

As we already discussed in detail, using complex passwords or passphrases is essential for system security. It's the lock on the front door of your user account and a foundational part of securing any system.

Encryption

Encryption is the process of scrambling readable data so that only authorized parties can view it. It's a foundational tool in cybersecurity that directly provides confidentiality, one of the three pillars of the CIA triad. By transforming your data into an unreadable format, encryption ensures it's kept secret from prying eyes. Furthermore, cryptographic systems that use encryption, such as digital signatures, also help provide integrity by proving that data hasn't been altered.

The Building Blocks of Encryption

The process uses a few key components to work its magic. The original readable message is called **plaintext.** A mathematical recipe known as a **cipher** (or algorithm), combined with a **key**, transforms the plaintext into scrambled, unreadable **ciphertext**. An authorized person then uses the same key and cipher to reverse the process and turn the ciphertext back into plaintext.

- **To Encrypt:** Plaintext + Cipher + Key = Ciphertext
- **To Decrypt:** Ciphertext + Cipher + Key = Plaintext

Securing Data at Rest & in Motion

Encryption is used to protect data in two primary states: when it's being stored and when it's being sent.

Data at Rest is data that's stored on a physical device. This includes files on your laptop's hard drive, your smartphone's storage, a server, or a USB thumb drive.

- **The Risk:** If the physical device is lost or stolen, an attacker has all the time in the world to try to access its data.

- **Encryption's Role:** Encrypting your data at rest is like storing your valuables in a high-tech safe. Even if a thief steals the entire safe (your laptop), they can't access its contents without the key (your password).

Data in Motion is data that's actively traveling from one place to another, such as across the Internet or a network.

- **The Risk:** The primary risk is that an attacker can eavesdrop on your connection and intercept your data as it travels, especially on untrusted networks like public Wi-Fi.
- **Encryption's Role:** Encrypting your data in motion is like putting a letter in a sealed, coded envelope before giving it to a messenger. Even if your data is intercepted, the thief only gets a scrambled message they can't read.

Understanding these two states is critical: encryption protects your data whether it's stored on your devices or traveling across a network.

The Two Flavors of Encryption

While we won't get into the deep technical details, it's helpful to know that there are two main forms of encryption: **symmetric** and **asymmetric**. The primary difference between them is how they use their keys, but both work to achieve the same goal of protecting your data.

As its name implies, symmetric encryption uses a single, shared secret key for both encrypting and decrypting. Because it's very fast, it's excellent for encrypting large amounts of data.

Asymmetric encryption, on the other hand, uses a pair of two different keys: a public key that can be shared with anyone, and a private key that you must keep secret. Because this form of encryption is slower, it's typically used for tasks that don't involve large amounts of data, such as creating digital signatures to verify identity.

In fact, most secure systems you use every day, like HTTPS websites, use both together. They use the slower asymmetric encryption at the beginning to securely exchange a secret key, and then switch to the faster symmetric encryption to protect your actual data.

> **GOING DEEPER: Not All Encryption is Created Equal**
>
> Just as with any technology, encryption algorithms can become outdated over time as computers become more powerful. Using a weak, outdated algorithm is like protecting your valuables with an old padlock that a modern bolt cutter can snap through in seconds. The current industry standard for strong encryption is AES (Advanced Encryption Standard). You should always ensure the tools and services you use employ AES. Older, "cracked" algorithms like DES, 3DES, and WEP (for wireless networks) are no longer secure and should never be used.

Common Use Cases for Encryption

You're already surrounded by encryption in both your personal and professional life. Here are a few common use cases you'll recognize.

Personal Use Cases

- **Secure Websites (HTTPS):** When you see the padlock icon in your browser's address bar, that means your connection to a website is encrypted. This protects sensitive information, such as your login credentials or credit card numbers, from being snooped on while it travels from your computer to the website's server.
- **Encrypted Messaging:** Popular apps like WhatsApp and Signal use end-to-end encryption to make sure that only you and the person you're talking to can read your messages. Not even the company that owns the app can see the content of your conversations, ensuring true confidentiality.
- **Securing Your Devices:** The full-disk encryption on your smartphone or laptop (like BitLocker for Windows or FileVault for Mac) uses encryption to protect all the files stored on it. If your device is lost or stolen, this makes your data completely unreadable to whoever finds it, safeguarding your data at rest.

Business Use Cases

- **Virtual Private Networks (VPNs)**: As we'll discuss later in this chapter, companies rely on VPNs to create an encrypted tunnel for employees working remotely. This allows them to securely access internal company resources from anywhere in the world as if they were in the office.
- **Database Encryption**: Businesses that store large amounts of sensitive customer data, such as personal information, health records, or financial details, use encryption to protect the entire database. If a hacker manages to breach the server, the encrypted data remains secure.

- **Secure Email**: While most email is encrypted in transit, some businesses use additional end-to-end encryption tools to protect highly confidential communications, such as legal documents or trade secrets, ensuring only the intended recipient can ever read the message.
- **Digital Signatures**: Businesses use digital signatures to provide non-repudiation for important documents and communications. A digital signature, created using asymmetric encryption, serves as a verifiable seal that proves who created or approved a document and confirms that it hasn't been altered since it was signed. This is crucial for contracts, financial approvals, and other documents that require legal proof of who authorized what.

! COMMON MISTAKE: Not Safeguarding Your Recovery Key

Encryption is powerful, but it can work against you if you lose your decryption key. When you first set up full-disk encryption like BitLocker or FileVault, the system generates a special recovery key, a long, unique code that serves as a master backup for your password. Imagine your computer's operating system crashes or the hard drive itself fails. To get your data back, you might move the drive to a new computer or restore your files from an encrypted backup.

In either scenario, you'll be prompted for your recovery key. In a corporate environment, the IT department often maintains a backup of your key for recovery purposes. But at home, you're the IT department. If you lose your recovery key, your data is permanently locked away forever. No one, not even a cybersecurity expert, will be able to get it back for you. Always make sure to print out your recovery key and store it in a safe, separate location.

Firewalls

A firewall is a fundamental network security device and a foundational component of a defense-in-depth strategy. Just about every network has a firewall, and many have more than one. In fact, your home network probably has multiple firewalls.

Figure 8.3: Firewall

Firewalls filter traffic between networks, allowing legitimate traffic while blocking unwanted or malicious connections. Think of a firewall as a security guard for your network, checking the ID of every data packet trying to enter or leave. Specifically, firewalls filter data packets as they pass through them based on a set of security rules you configure. Based on these rules, firewalls then determine which packets are allowed into your network and which are blocked.

To make these decisions, firewalls examine information in the packet headers, such as the Source and Destination IP addresses (where the packet came from and where it's going), the Port numbers (what application it's for), and the Protocol (TCP, UDP, etc.). You might configure a firewall rule that says, "Allow incoming traffic on port 443 (HTTPS) to our web server, but block all other incoming traffic." This lets customers access your website while blocking all other traffic.

Types of Firewalls

While all firewalls filter traffic, they do so with different levels of intelligence. It's important to understand the different approaches they use.

- **Packet-Filtering Firewalls**: These are the most basic type of firewall. They look at each data packet in isolation and make a simple allow-or-deny decision based on a pre-defined set of rules that check its IP address and port number. While very fast, they don't understand the traffic's context.
- **Stateful Firewalls**: These are a major step up. Stateful firewalls monitor and track the state of active connections. This allows them to understand the context of a conversation and automatically determine whether a packet is part of a legitimate, ongoing network session initiated from inside the network. This is the most common type of firewall in use today.
- **Application Layer Firewalls**: Also known as proxy firewalls, these advanced firewalls operate at the application level and act as intermediaries (proxies) for your traffic. This allows them to inspect the actual content of your traffic, not just the headers. You can create very granular rules to filter what is and isn't allowed. For example, they can distinguish between legitimate web requests and web-based attacks or block access to specific websites.
- **Next-Generation Firewalls (NGFWs)**: NGFWs are the modern standard, combining all the features of the firewalls above, packet filtering, stateful inspection, and application-layer awareness, with additional security services. They often include an integrated Intrusion Prevention System (IPS) to block known attacks, provide deep application awareness to see exactly which programs are generating traffic, and have the ability to use threat intelligence feeds to block the latest known threats from the Internet.

Network-Based vs. Host-Based Firewalls

Firewalls are deployed in two primary locations, each with a distinct purpose. A network-based firewall is designed to protect either the entire network (a perimeter firewall) or internal network segments, while a host-based firewall protects the individual system it's installed on. Using both together is a key part of a defense-in-depth strategy.

- **Network-Based Firewall**: It's typically placed at the network's perimeter (the edge where the internal network connects to the Internet) to inspect all traffic entering and leaving the network, serving as the first line of defense for every device on that network. Larger networks also commonly have additional network-based firewalls protecting network segments.
- **Host-Based Firewall**: As we touched on in the "Endpoint Protection" section, a host-based firewall is software that runs on an individual device (or host). It monitors and controls traffic for that specific device only, protecting it from threats that might already be inside the network. Most modern operating systems, such as Windows and macOS, include built-in firewalls.

This layering of firewalls is a perfect example of defense in depth. Your home network probably has several. Your wireless router acts as your network firewall, and your laptop or desktop also has its own host-based firewall.

In a business environment, an organization typically installs a perimeter network-based hardware firewall to protect its entire internal LAN from the Internet. They can also place additional network firewalls within the network to create secure segments. For example, separating the finance department's network from the engineering network. On top of that, they'll install host-based firewalls on all systems on the network. While firewalls aren't going to stop 100% of attacks, they're your first line of defense and will stop most of them.

> **COMMON MISTAKE**: Disabling your computer's built-in firewall because you have a network firewall. Network firewalls protect at the perimeter, while host-based firewalls protect individual devices from threats that get past the perimeter or originate within the network (such as an infected laptop brought from home). Both provide essential protection at different layers.

Intrusion Detection & Prevention Systems (IDS/IPS)

While firewalls filter traffic based on rules, intrusion detection systems and intrusion prevention systems (IDS/IPS) act as detectives looking for suspicious behavior. These are considered more advanced network security devices. Firewalls are excellent at blocking traffic based on rules you define, but what about sophisticated attacks that mimic legitimate traffic? That's where IDS/IPS comes in. While a firewall acts like a security guard with a list of approved guests, an IDS/IPS acts like a detective, looking for suspicious behavior.

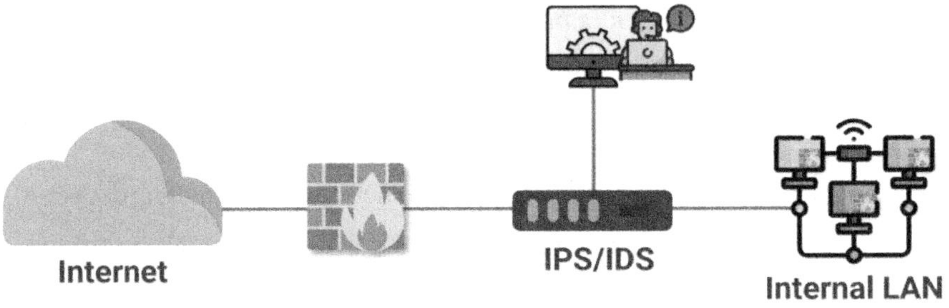

Figure 8.4: IDS/IPS System

An IDS or IPS is designed to detect attacks. Based on whether it's an IDS or an IPS, it'll respond either passively or actively.

KEY POINT: Firewalls block based on rules you define, meaning they need you to tell them what to block. IDS/IPS systems detect attacks based on known attack patterns (signatures) and suspicious behavior (anomalies), even if you haven't written specific rules for them. This means they can catch threats you didn't even know existed.

Intrusion Detection Systems (IDS)

The intrusion detection system is passive, meaning it responds by logging an event and notifying IT staff, and does nothing beyond that. Let's say a black hat hacker on the Internet tries to send a malicious packet of data into your internal network. It gets past your firewall because it appears to be legitimate traffic on an allowed port, passes through your IDS, and then gets into your protected internal local area network. The IDS recognizes the packet's behavior as suspicious and logs the activity, sending an alert to a monitoring station where the IT staff determines how to mitigate the issue.

Intrusion Prevention Systems (IPS)

The intrusion prevention system, on the other hand, responds actively, meaning it changes something within the network environment to stop the attack, such as changing an access control list rule, closing a process or session, or even blocking a port. It does what needs to be done to stop that attack. Using the same example, when the hacker tries to send malicious data into your network, and it gets past your firewall to the IPS, the IPS recognizes the attack pattern and actively stops that attack. For example, it might change an access control list rule, log the event, and send it to the management station, where IT staff are notified.

Why Use an IDS Instead of an IPS?

You may be asking yourself, "Why would an organization ever use an IDS if it could use an IPS that stops the attack?" It all comes down to the criticality of the network.

In certain organizations, we don't want any network changes to occur without consulting IT managers and network engineers because a change may have unintended consequences, breaking something. So in certain instances, an IDS is preferred over an IPS. It all comes down to the network design, the organization, and how they want to manage network changes. That's why we would sometimes use the more passive method of logging and notifying IT staff with an IDS, and in other instances, we'd use an IPS and be okay with it changing the network environment on its own.

A hospital might use an IDS instead of an IPS for systems connected to life-support equipment. They can't risk an IPS automatically blocking traffic that might be legitimate but looks suspicious; a false positive could have life-threatening consequences. They prefer human review before taking action.

IDS/IPS Detection Methods

IDS/IPS systems use two primary detection methods:

- **Signature-Based Detection**: Matches traffic against known attack patterns (like anti-malware signatures). Very accurate for known threats, but can't detect new attacks.
- **Anomaly-Based Detection**: Establishes a baseline of normal network behavior, then flags deviations. It can detect zero-day attacks but generates more false positives.

Most modern systems use both methods together for comprehensive protection.

Using a Network DMZ

A DMZ (Demilitarized Zone), also called a screened subnet, is a perimeter network designed to be securely separated from an organization's private internal network. It allows untrusted users on the Internet to access specific public services, such as your company website, while blocking them from accessing your internal network, which houses sensitive data and business systems. The whole goal of a DMZ is to solve a critical problem: how do you provide public-facing services that anyone on the Internet can access while simultaneously protecting your private, trusted internal network, which houses your sensitive business data and systems? The DMZ provides both public accessibility and internal security. Think of it as a secure waiting room between the dangerous outside world and your valuable internal resources.

How a DMZ is Configured

A DMZ can be implemented in two ways, both using firewalls to create the isolated zone:

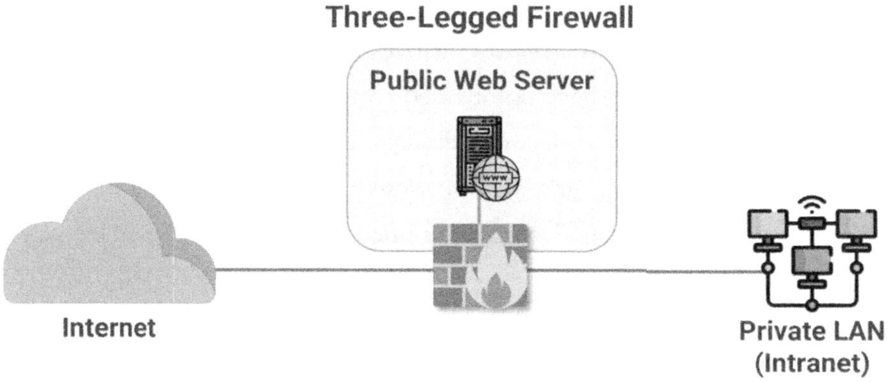

Figure 8.5: DMZ Three-Legged Firewall Configuration

Three-Legged Firewall Configuration: A single firewall with three network interfaces: one connected to the Internet, one to the DMZ, and one to your internal network. The firewall enforces strict rules about what traffic can flow between these three zones. This is the most common and cost-effective approach for small to medium businesses.

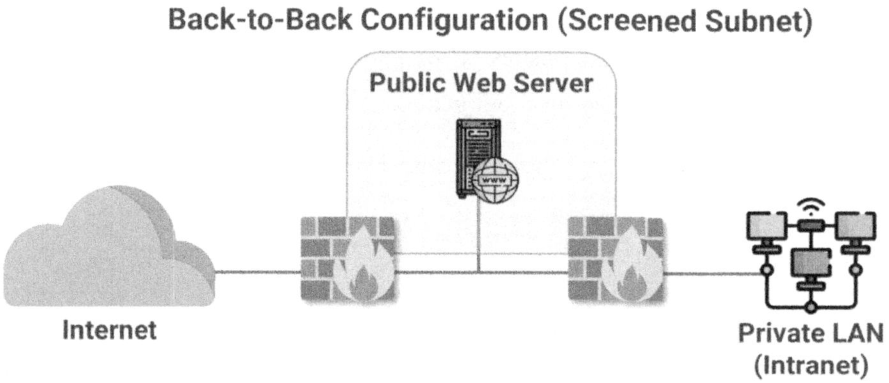

Figure 8.6: DMZ Back-to-Back Firewall Configuration (Screened Subnet)

Back-to-Back Firewall Configuration (Screened Subnet): Two separate firewalls with the DMZ sandwiched between them. The first firewall sits between the Internet and the DMZ, while the second firewall sits between the DMZ and your internal network. This provides stronger security because an attacker must compromise two separate firewalls to reach your internal network. Larger enterprises and high-security organizations typically use this configuration despite the higher cost and complexity.

Think of it like a bank: customers are allowed in the public lobby to conduct transactions at teller windows, but they can't access the vault or back offices where cash, records, and sensitive systems are stored without passing through another layer of security–reinforced doors and employee-only access. Looking at figures 8.5 and 8.6, you can see how this same separation is created by the firewall (or firewalls), controlling exactly what traffic flows between the Internet, the DMZ, and your internal network.

What Goes in a DMZ?

So what types of services actually belong in a DMZ? Organizations typically place any public-facing services that external users need to access, including:

- **Public website** (web server)
- **File server** for software downloads or drivers (FTP server)
- **Public email server** (if you were an email provider like Gmail)
- **Private portal** for business partners (Private portal for authorized suppliers or customers)
- **VPN concentrators** for remote access
- **DNS servers** that respond to external queries

If an attacker compromises your public web server in the DMZ, they've only breached the DMZ, not your internal network, where sensitive data, employee workstations, and business systems reside. The firewall between the DMZ and the internal network provides a critical second layer of defense.

Consider a retail company with a public e-commerce website placed in the company's DMZ, enabling customers to browse products and make purchases. When a customer places an order, the web server in the DMZ sends a limited query to the internal database server to check inventory and process payment. However, customers never directly access the internal network where employee data, financial records, and business systems are stored. Therefore, if a hacker compromises the web server, they're stuck in the DMZ and can't pivot directly to the internal network where the real treasure trove of data resides.

Honeypot Decoys

Imagine setting up a fake, poorly secured safe filled with fake money to catch a thief. That's essentially what a honeypot does in cybersecurity.

Figure 8.7: Honeypot Decoy System

A honeypot is an intentionally vulnerable decoy system designed to attract attackers. Organizations typically place honeypots in the DMZ or scattered throughout the internal network. They look like legitimate, valuable targets, perhaps a server named PAYROLL-BACKUP or CUSTOMER-DATABASE, but they contain no real data or business systems. Security teams intentionally make honeypots appear attractive by using outdated software, weak passwords, and names that sound valuable.

The Purpose of Honeypots

Honeypots serve multiple strategic purposes:

- **Distraction**: They distract attackers by luring them away from real systems toward the decoy.
- **Detection**: They provide instant detection, alerting security teams the moment someone accesses the honeypot, since there's no legitimate reason anyone should be there.
- **Intelligence**: They gather intelligence by allowing security teams to observe attack techniques, tools, and methods in a controlled environment without risk to real systems.
- **Time-Wasting**: They waste attackers' time, keeping them busy while security teams respond and investigate.

Honeypots function as both an early warning system and a learning laboratory. They allow security teams to detect intrusions immediately while gaining valuable intelligence about attacker behavior—what tools they use, which vulnerabilities they exploit, and which data they're targeting—all without putting real systems and data at risk.

Honeynets: Taking It Further

Organizations can take honeypots a step further by deploying a honeynet: a network of multiple interconnected honeypots that simulates an entire realistic environment. Honeynets mimic a complete network infrastructure, including servers, workstations, and user activity. This provides even deeper insights into the sophisticated behavior of advanced attackers and reveals the complex tactics and techniques they employ.

Virtual Private Network (VPN)

So far, we've focused on protecting the perimeter of our network with firewalls, detecting threats with IDS/IPS, and isolating public services in a DMZ. But what happens when employees need to work from home or travel? How do they securely access company resources across the Internet? This is where virtual private networks (VPNs) come in. A VPN creates a secure, encrypted connection between our device and a VPN server, protecting our data as it travels across untrusted networks like public Wi-Fi or the Internet.

Understanding How VPNs Work

Imagine you're a famous celebrity and the paparazzi are constantly on your trail, trying to snap pictures, see where you go, and eavesdrop on your conversations. On the Internet, attackers and snoopers can act like digital paparazzi, trying to follow your every move and intercept your data.

A VPN is your solution to this problem. Think of it like having access to a private, underground tunnel. This tunnel takes you directly from your house (your computer) to your destination (a website or your office network), and no one on the outside can see you, track where you're going, or know what you're carrying. Even if someone is watching the entrance and exit of the tunnel, they can't see inside or know what's happening within it.

This is how VPNs work: they use what's called a tunneling protocol to wrap your data in a protective, encrypted layer before sending it across the Internet. Even if someone intercepts your data while it's traveling, all they see is scrambled, unreadable information. Only you and the VPN server can see what's actually inside that secure tunnel.

The Three Primary Use Cases

Now that you understand the basics of how a VPN works, we'll explore its three primary use cases. VPNs serve different purposes depending on who's using them and why. We'll look at the three main scenarios you'll encounter.

1. Remote Access VPN (Corporate)

A remote access VPN allows employees to securely connect to their company's internal network from remote locations. When you work from home, your company has installed VPN client software on your work laptop. When you launch the VPN client and connect, it establishes an encrypted tunnel from your home computer through the Internet to a VPN server at your company's office. Once connected, your computer functions as if it's physically plugged into the office network. You can access file servers, internal applications, databases, and other resources just as you would if you were sitting at your desk in the office. All your traffic flows through this encrypted tunnel, protecting sensitive company data from being intercepted on your home Wi-Fi or Internet connection.

Figure 8.8: Remote Access VPN (Corporate)

For example, let's assume you're working from your home office and need to access a confidential client file stored on your company's internal file server. You launch the VPN client on your laptop and connect to your company's network. Even though you're miles away from the office, you can now access the file server, open the document, and work on it securely. Your neighbors using the same Internet service provider can't see what you're accessing, and your ISP can only see encrypted traffic going to your company; they can't see that you're viewing confidential client files.

2. Site-to-Site VPN (Corporate Network Connections)

A site-to-site VPN connects entire office locations to each other, allowing multiple sites to function as a single unified network. Instead of individual employees connecting via VPN client software, the company installs VPN gateway devices at each physical location, such as its headquarters, branch offices, and data centers. These gateways automatically create a permanent, encrypted tunnel between each location, providing a cost-effective way to securely connect the offices over the public Internet. This allows employees at any location to access resources at any other location as if they were on the same local network, even though the sites might be in different cities or countries.

Figure 8.9: Site-to-Site VPN (Corporate Network Connections)

For example, let's assume a company has its headquarters in New York and a branch office in Los Angeles. Instead of having LA employees connect individually via remote access VPNs, the company installs VPN gateways at both locations. These gateways create a permanent, encrypted tunnel between New York and LA. Now, when an LA employee needs to access the accounting database in New York, they simply open it as if the server were in their own office. When a New York employee needs to print a document on the LA office printer, they select it as they would any other printer on the network. The encryption and tunnel happen automatically in the background, and the two offices operate as one seamless network.

3. Privacy VPN (Consumer Services)

The last primary use case we'll look at is privacy VPNs. These are the popular consumer services you've likely seen advertised, such as NordVPN, ExpressVPN, or ProtonVPN. If you use a VPN at home or for personal travel, this is most likely the type you're using. They're designed for individual consumers who want to protect their privacy and security when using the Internet, especially on untrusted networks like public Wi-Fi. When you connect to a privacy VPN, your device creates an encrypted tunnel to the VPN provider's server. Your Internet traffic then flows through this tunnel to the VPN server, which forwards your requests to websites on your behalf. This hides your Internet activity from your Internet service provider, from the owner of the public Wi-Fi network you're using, and from anyone else monitoring your connection. Websites you visit see the VPN server's location instead of your actual location, which also allows you to access content that might be restricted in your region.

Figure 8.10: Privacy VPN (Consumer Services)

For example, imagine you're on vacation on a cruise ship and connect to the ship's public Wi-Fi. Before you connect to your privacy VPN service, any website you visit is visible to the network provider. After you connect to your VPN, even though you're on the public Wi-Fi (which could be monitored by other guests or even malicious actors), all your Internet activity is encrypted inside the VPN tunnel. The cruise ship's network can only see that you're sending encrypted data to the VPN provider—they can't see that you're checking your bank account, scrolling through Facebook, or reading your Gmail. Your personal data is protected from eavesdropping.

It's also important to understand that VPNs only encrypt traffic inside the tunnel (from your device to the VPN server). Once traffic leaves the VPN server and reaches a website, the website handles encryption using HTTPS. This is why you should always look for the padlock icon for HTTPS websites, even when using a VPN, to ensure the entire connection is secure from end to end.

! **COMMON MISTAKE**: Thinking a VPN makes you completely anonymous online. VPNs hide your activity from your ISP and local network, but the VPN provider can see everything you do. Therefore, choosing a reputable VPN provider with strong privacy policies and a no-logging commitment is essential. You'll find that free VPN services often monetize by selling your browsing data.

It's important to understand that a VPN serves a different purpose than the other network defenses we've discussed, such as firewalls, IDSs, and DMZs. While those countermeasures are designed to protect the perimeter of your private network from external attacks, a VPN is designed to protect your communications as they travel across the untrusted Internet. Whether you're a remote employee accessing company files or a traveler using public Wi-Fi at an airport, a VPN's primary job is to provide a secure, private connection, ensuring no one can eavesdrop on your data while it's in transit.

Wireless Network Security

The final foundational countermeasure we'll explore in this chapter addresses a technology that's become ubiquitous in our homes, offices, and public spaces: wireless networks. While wireless connectivity offers tremendous convenience, it also introduces unique security challenges that require specific protections.

Understanding the Wireless Vulnerability

Wired networks are like having a private, whispered conversation in a soundproof room; you have to be physically connected to the wire to "hear" the data traveling through it. An attacker would need physical access to the cable itself, which is usually locked away in walls, ceilings, or secure network closets. Wireless networks, on the other hand, are like shouting your conversation across a crowded room because anyone nearby with the right equipment can potentially listen in.

The inherent vulnerability of wireless networks stems from a simple fact: they transmit data through the air via radio signals. These signals don't stop at your walls or property line. They emanate outward in all directions, and anyone within range with a wireless adapter can detect your network and potentially intercept the traffic flowing across it.

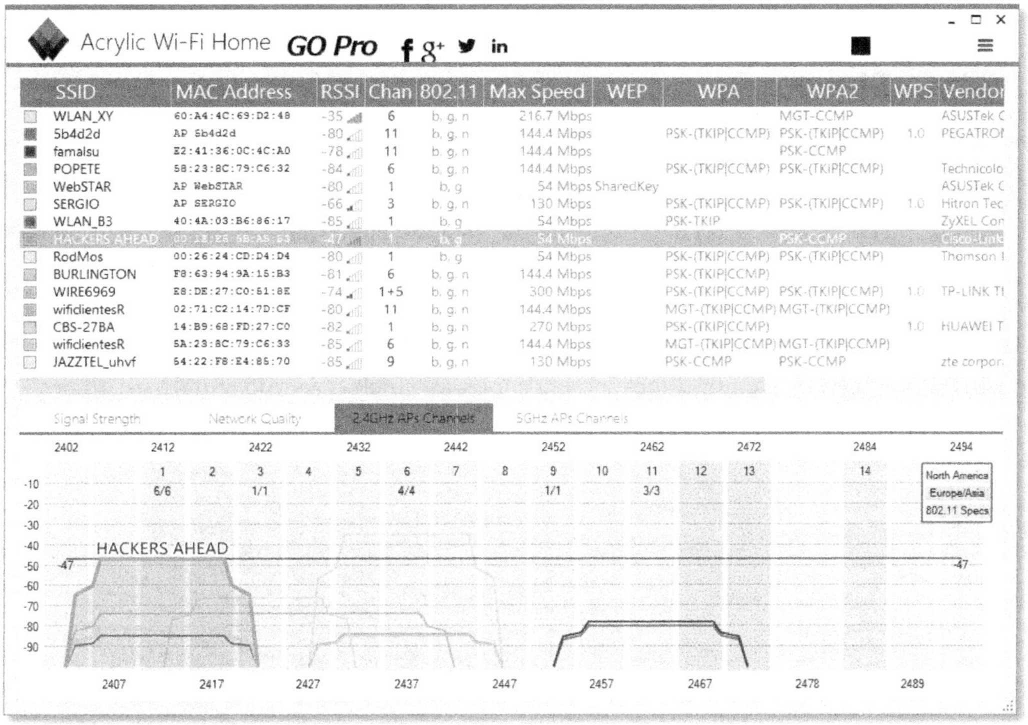

Figure 8.11: Wireless Network Scanning Software

Think about what connects to your wireless network: your laptop with work documents, your smartphone with personal emails and banking apps, your smart TV, your home security cameras, maybe even your thermostat and door locks. All of these devices are transmitting data through the air, and without proper security, an attacker sitting in a car outside your home or in the coffee shop across the street could intercept this traffic.

In a business environment, the stakes are even higher. Wireless networks often provide access to the same internal resources as wired networks: file servers, databases, and sensitive business systems. If your wireless network isn't properly secured, an attacker doesn't need to physically break into your building or plug into your network; they can simply launch an attack from the parking lot.

Here's something that might surprise you: discovering wireless networks and learning detailed information about them is trivially easy. Free software tools that anyone can download allow attackers to see every wireless network in range and gather extensive information, including what type of encryption you're using and whether you have known vulnerabilities. This ease of reconnaissance is exactly why wireless network security is so critical.

Wireless Encryption Standards

The primary defense for wireless networks is encryption. Remember from earlier in this chapter how encryption scrambles data so that even if someone intercepts it, they can't read it? That same principle applies to wireless networks. When you enable wireless encryption, all the data traveling through the air between your devices and your wireless access point is scrambled, making it unreadable to anyone trying to eavesdrop.

However, not all wireless encryption is created equal. Over the years, several wireless security protocols have been developed, and some have been broken by attackers. Let's look at which ones you should use today.

Figure 8.12: Wireless Security Settings

WEP & WPA: Never Use These

WEP (Wired Equivalent Privacy) was the first wireless encryption standard, introduced in 1997. The goal was to provide wireless networks with a security "equivalent" to wired networks, but the implementation was fatally flawed. Although weaknesses had been known earlier, in 2005 the FBI publicly demonstrated that WEP could be cracked in mere minutes using freely available tools.

WPA (Wi-Fi Protected Access) was introduced in 2003 as an interim solution while the industry worked on a more robust standard. Although better than WEP, it still had known vulnerabilities and has since been superseded.

> ⚠ **COMMON MISTAKE: Thinking WEP is "better than nothing"**
>
> WEP provides a false sense of security. It's so easily broken that it offers essentially no protection against anyone with basic technical knowledge. Using WEP is like storing cash in a lock box where the key is taped to the lid—it won't stop anyone who actually wants to get in.

WPA2 & WPA3: Use These

WPA2 (Wi-Fi Protected Access 2), introduced in 2004, uses the Advanced Encryption Standard (AES), the same strong encryption algorithm used by governments and militaries around the world. When properly configured with a strong password, WPA2 provides solid security for wireless networks. For nearly two decades, WPA2 has been the security standard and is supported by virtually all wireless devices manufactured since 2006. It's secure enough for most home and business uses, provided you follow critical guidelines, such as using a strong passphrase and disabling WPS (Wi-Fi Protected Setup).

WPA3, introduced in 2018, is the latest wireless security protocol and provides stronger encryption and better protection against password-guessing attacks than WPA2. As of 2025, WPA3 support has become widespread in newer devices, though you may still encounter some older devices (typically those manufactured before 2019–2020) that don't support it. You can only use WPA3 if both your wireless router and all your devices support it, so if you have any older devices on your network, you may need to use WPA2 instead.

If you're buying a new wireless router today, choose one that supports WPA3, as most routers manufactured in the last few years include it. Once you have WPA3 available, enable it if all your devices support it for the strongest protection. However, if you have older devices that don't support it, WPA2 remains a strong alternative. And remember, never use WEP or WPA under any circumstances.

Personal vs. Enterprise Mode

When you configure wireless security on your router, you'll notice options for "Personal" and "Enterprise" modes. Let's clarify what these mean.

Personal Mode is what you use at home and in small businesses. Everyone who connects to the network uses the same password. When you give your Wi-Fi password to a house guest, you're giving them the same password that every device on your network uses. This is easy to set up and perfectly adequate for home and small-business use.

Enterprise Mode uses **802.1x authentication** and is what large organizations use. Instead of everyone sharing the same password, each user authenticates with their own individual username and password. This provides individual accountability (you know exactly who's on the network) and makes it easy to revoke access (when an employee leaves, IT simply disables their account).

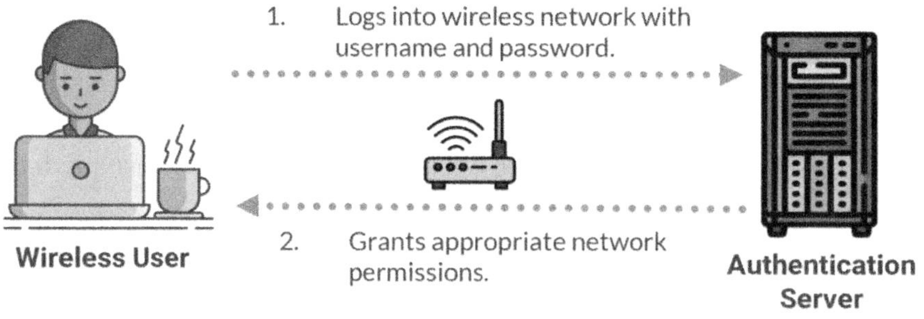

Figure 8.13: 802.1x Authentication

> **EXAMPLE**: A university uses 802.1x authentication for campus Wi-Fi. Each student logs in with their university credentials. When a student graduates, IT disables their account, and they immediately lose Wi-Fi access. If the university used a shared password instead, they'd have to change it and redistribute it to thousands of remaining students every time access needed to be revoked.

Common Wireless Security Mistakes

Now that you understand wireless encryption standards, we turn to some of the most common wireless security mistakes that leave networks vulnerable to attack, and how to avoid them.

- **Using Default Admin Passwords**: Every wireless router ships with a default administrator password, which is often admin/admin or admin/password. These are published online and are the first thing an attacker will try. You should change the administrator password immediately when you first set up a new router.
- **WPS Enabled**: Wi-Fi Protected Setup (WPS) was designed for convenience, but has serious security vulnerabilities that might allow attackers to crack your Wi-Fi password. You should disable WPS in your router settings because the convenience isn't worth the risk.

- **Using Weak Encryption**: As you already know, WEP can be cracked in minutes using freely available tools, and the original WPA has known vulnerabilities that make it unsafe. You should use only WPA2 or WPA3 encryption, and never select WEP or WPA in your router's settings.
- **No Guest Network**: When friends or family visit, giving them your main network password gives their devices access to everything on your network—your computers, printers, and smart home devices. You should create a separate guest network that provides Internet access but isolates visitors from your personal devices and files. Most modern routers make this easy to set up.
- **Running Outdated Firmware**: Your router is a computer running software (called firmware), and like any software, it can have security vulnerabilities. Router manufacturers regularly release firmware updates that fix newly discovered security flaws, but these updates often don't install automatically. You should log in to your router's administration page every few months to check for firmware updates.
- **Broadcasting Identifying Information**: Avoid including personal information when naming your wireless network. For example, a network named "Smith Family" tells attackers exactly who you are and where you live, making you an easier target for social engineering or physical attacks.
- **Using a weak Wi-Fi password**: Your Wi-Fi password is the key to your entire network. You should use a password of at least 12–15 characters, or better yet, a long passphrase made up of random words. A weak password like `password123` can be cracked through brute-force attacks in hours or days, while a strong passphrase would take centuries to crack.

These wireless security principles aren't just theoretical best practices; they have real-world consequences when ignored, as the following case study dramatically illustrates.

Case Study: TJX Companies Data Breach

Background

Between 2005 and 2007, TJX Companies, the parent company of retail brands including T.J. Maxx, Marshalls, and HomeGoods, suffered one of the largest retail data breaches at the time. Over an 18-month period, hackers stole more than 45 million credit and debit card numbers from TJX systems, compromising customer data across hundreds of stores. The breach revealed fundamental failures in how the company secured its wireless networks.

The Attack

The breach began with a simple but devastating weakness: TJX used Wired Equivalent Privacy (WEP) to secure its wireless networks in its stores. WEP had been publicly known as broken and insecure since 2001, and by 2004, it had been officially deprecated. The industry had moved on to WPA in 2003, yet TJX continued using the outdated, easily crackable encryption.

The attackers didn't need sophisticated tools or techniques. They simply sat in the parking lots of TJX stores with laptops and used freely available software to crack the WEP encryption on the wireless networks that transmitted data between price-checking devices, cash registers, and store computers. Breaking through the WEP encryption took only minutes.

Once through the weak wireless encryption, the attackers could see all the unencrypted traffic flowing across the network. They intercepted employee login credentials as employees accessed the company's systems, then used these stolen credentials to create their own accounts that appeared legitimate. With this access established, attackers operated undetected for over 18 months, systematically collecting transaction data and stealing millions of credit card numbers from customer purchases.

Impact

The breach ultimately cost TJX an estimated $256 million in lawsuits, settlements, and security remediation expenses. The company faced lawsuits from banks whose customers were affected, as well as from customers whose data was stolen. Beyond the financial cost, TJX's reputation suffered significant damage because customers lost trust in the company's ability to protect their payment information.

What makes this breach particularly significant is that it was entirely preventable. Court documents and security investigations revealed that TJX was aware that its wireless networks used outdated encryption prior to the breach. The cost of replacing WEP with WPA across their stores would have been a fraction of the $256 million they ultimately paid.

Lessons Learned

- **Use Current Encryption Standards, Not Deprecated Ones**: WEP was already known to be broken when TJX's breach occurred. Using known-broken security technology like WEP isn't just poor practice—it's a critical security failure. Organizations should keep wireless security up to date and upgrade to WPA2 or WPA3.

- **Wireless Signals Don't Stop at Your Walls**: The attackers conducted their reconnaissance and attacks from parking lots, never needing to enter TJX stores physically. This demonstrates why wireless security is so critical; your network is accessible to anyone within signal range.
- **The Cost of Proper Security Is Far Less than the Cost of a Breach**: Upgrading from WEP to WPA would have cost TJX a relatively small amount compared to the $256 million the breach ultimately cost them. Investing in proper wireless security isn't an expense—it's protection against catastrophic losses.
- **Knowing About Vulnerabilities Without Fixing Them Is Unacceptable**: TJX was aware that its wireless encryption was inadequate, but failed to act. When you know about a security weakness, especially one as fundamental as using broken encryption, you should prioritize fixing it immediately.

Bringing It All Together

This chapter presented the fundamental countermeasures that form the foundation of a defense-in-depth strategy. Each security control addresses different attack vectors, and together they create overlapping layers of protection that make successful attacks significantly more difficult.

We saw how multi-factor authentication dramatically reduces credential theft by requiring multiple forms of verification. We learned that password managers and passphrases solve the problem of creating and remembering strong, unique passwords for every account. And we covered how endpoint protection through system hardening closes potential entry points by disabling unnecessary services, removing unused accounts, and keeping systems patched.

Encryption protects data confidentiality whether it's stored on devices or traveling across networks. Even if attackers intercept encrypted data, they can't read it without the decryption key. Network security controls create defensive perimeters through firewalls that filter traffic, IDS/IPS systems that detect attack patterns, and DMZs that isolate public services from internal networks.

VPNs create encrypted tunnels for secure remote access and protect your data on untrusted networks. Wireless security requires current encryption protocols, WPA2 or WPA3, to prevent attackers from intercepting traffic transmitted over the air. The TJX breach demonstrated the catastrophic consequences of using outdated WEP encryption, resulting in 45 million stolen credit card numbers and $256 million in costs.

Defense in depth means no single security failure results in complete compromise. Layered security ensures that attackers must overcome multiple barriers, increasing the likelihood of detection and prevention.

Key Takeaways

- **Multi-factor authentication (MFA) dramatically reduces credential theft** by requiring at least two of the following factors: something you know, something you have, and something you are.
- **Password managers solve the problem of remembering dozens of unique passwords**; you only need to remember one strong master password to access all of them.
- **Endpoint protection through system hardening reduces attack surfaces** by disabling unused ports and services, removing unnecessary accounts, keeping systems patched, and using anti-malware software.
- **Encryption protects data at rest and in motion**; even if attackers intercept it, they can't read it without the decryption key.
- **Firewalls filter traffic while IDS/IPS systems detect attacks**. IDS alerts administrators to threats while IPS actively blocks them.
- **VPNs create encrypted tunnels** that are essential for remote work and for protecting data on untrusted networks, such as public Wi-Fi.
- **Wireless networks require WPA2 or WPA3 encryption**; remember, deprecated protocols like WEP can be cracked in minutes, as demonstrated by the TJX breach that compromised 45 million credit cards.
- **Defense in depth implements multiple overlapping security layers** so that no single security failure results in complete compromise.

🔧 Quick Action: Strengthen Your Digital Defenses

In this chapter, you learned about fundamental threat countermeasures, including multi-factor authentication, password management, endpoint protection, encryption, firewalls, VPNs, and wireless security. So in this quick action exercise, you'll complete three essential security checks that address the most common vulnerabilities in personal accounts and home networks.

What to Do

Enable Multi-Factor Authentication (MFA) on Critical Accounts:

- Identify your most important accounts: primary email, banking, investment accounts, and major online services.
- Log in to each account and enable MFA in the security settings.

Consider Using a Free Password Manager:

- Many reputable password managers offer free versions, such as Bitwarden and Proton Pass.
- Research them and consider giving one a try.

Verify Your Home Wi-Fi Security:

- Confirm your Wi-Fi uses WPA2 or WPA3 encryption.
- Change the default admin password if you haven't already.
- Disable WPS (Wi-Fi Protected Setup) if the option exists.
- Check for and install any available firmware updates.

By completing these three checks, you've significantly strengthened the security of your personal accounts and home network, making yourself a much harder target for attackers.

Chapter 9. Security Operations

Introduction

In the previous chapters, we've explored the core concepts of cybersecurity, including understanding threats, vulnerabilities, and fundamental countermeasures. We've discussed firewalls, encryption, access controls, and other defensive technologies that form the foundation of a strong security posture. But here are three critical questions: Once an organization has implemented these security measures, how does it know they're working? How do they detect when something goes wrong? And when an incident occurs, how do they respond effectively?

This is where security operations comes into play. Security operations is the ongoing practice of monitoring, detecting, analyzing, and responding to security events in real-time. It's the difference between simply having security tools installed and actually using them effectively to protect your organization. Think of it this way: you can install the best alarm system in your home, but if no one's monitoring it or responding when it goes off, what's the point?

> **DEFINITION: Security operations** is the ongoing practice of monitoring, detecting, analyzing, and responding to security events in real-time.

In this chapter, we'll explore the key components of security operations, including continuous monitoring, endpoint detection and response (EDR), incident response, and proactive security assessments. And to bring these concepts to life, we'll analyze one of the most significant security breaches in retail history, the 2013 Target breach, to understand what happens when security operations fail and what lessons we can learn.

Case Study: The Target Breach

Background

In November 2013, cybercriminals infiltrated Target's network and installed malware on point-of-sale (POS) systems across the retail chain. Over several weeks during the peak holiday shopping season, the malware captured credit and debit card information from approximately 40 million customers, along with personal information from an additional 70 million individuals. The breach became one of the most significant retail security incidents in history and a defining example of security operations failure.

The Attack

The attackers didn't breach Target's network directly. Instead, they compromised Fazio Mechanical Services, a third-party HVAC vendor with network access to Target's systems for billing and project management. After stealing the vendor's credentials, attackers used them as an entry point into Target's network.

Once inside, the attackers moved laterally through Target's network until they reached systems controlling POS terminals in stores nationwide. They installed "BlackPOS" malware specifically designed to capture payment card data from memory before encryption. The stolen data was collected on compromised servers within Target's network, then transmitted to external servers controlled by the attackers, often during peak shopping hours when the traffic would be less noticeable.

The Security Operations Failure

Target had invested $1.6 million in FireEye malware detection systems, which detected the malware multiple times and alerted Target's security team. However, despite receiving multiple alerts, Target's security team didn't act on these warnings. Additionally, Target had disabled the FireEye system's automatic malware-deletion feature, opting for human review, but that review never led to any action. The breach continued undetected for weeks until Target was notified by external parties, specifically, the Department of Justice and banking industry sources, who noticed unusual fraudulent transaction patterns on customer cards.

Impact

Target ended up paying $18.5 million to settle with 47 states and the District of Columbia, plus additional settlements to banks and credit card companies. The company's CEO and CIO both resigned. And beyond financial costs, Target suffered severe reputational damage and lost customer trust.

Lessons Learned

- **Detection Without Response Is Meaningless**: Target's security tools detected the breach, but the alerts weren't acted upon. Organizations should have effective processes to ensure alerts are reviewed, investigated, and responded to promptly.

- **Third-Party Risk Must Be Managed**: The breach originated through a vendor with network access. Security operations should effectively monitor third-party access and ensure vendors meet security standards.
- **Network Segmentation Limits Breach Scope**: Proper segmentation may have prevented attackers from moving from the HVAC systems to the payment systems. If Target's billing and project management systems were properly isolated from their POS systems, the hackers might never have gained access to them.
- **Alert Fatigue Requires Management**: Target preferred manual human review over the FireEye system's automatic malware-deletion feature, which may have led to alert fatigue, leaving their security team overwhelmed by alerts and potentially missing critical warnings. Unfortunately, this played a significant part in the breach going undetected.

The Target breach is a powerful lesson in what happens when security operations break down. The tools were in place, and the alerts were firing, but no one was truly watching and acting on them. This case study highlights why organizations need a formal, ongoing practice to ensure threats aren't only detected but also responded to promptly. The foundation of that practice is called continuous monitoring, which we'll explore next.

Continuous Monitoring

Continuous monitoring is the practice of maintaining ongoing awareness of your organization's security posture, vulnerabilities, and threats. Rather than conducting periodic security checks, such as once a week, bi-weekly, or even quarterly, continuous monitoring means you're always watching, always analyzing, and always ready to respond. Think of continuous monitoring like a security guard who never takes a break. Traditional security approaches were more like having a security guard who checks the building once a day. By the time they discover something's wrong, the damage may already be done. Continuous monitoring ensures that you're aware of security events as they happen, not hours or days later.

Why Continuous Monitoring Matters

The threat landscape moves fast. Attackers don't work on a 9-to-5 schedule, and they don't wait for convenient times to strike. A vulnerability that was unknown yesterday might be discovered and exploited today. An employee might click on a phishing link at 2:00 am. A piece of malware might start exfiltrating data during the weekend when fewer people are watching. Continuous monitoring addresses these realities by providing real-time or near-real-time visibility into what's happening across your environment.

KEY POINT: The sooner you know about a problem, the sooner you can stop it and the less damage occurs. Studies have shown that the average time between when an attacker gains access to a network and when they're discovered, called "dwell time," can be weeks or even months. During that time, attackers can steal data, install backdoors, and cause extensive damage, which you've seen from our numerous examples and case studies throughout the book. Continuous monitoring helps reduce this dwell time dramatically, sometimes from months to minutes.

What Gets Monitored

So what exactly are we monitoring? Effective continuous monitoring covers multiple layers of your environment. Let's break this down in a way that makes sense.

The Network

First, think about your network, all the data flowing in and out of your organization. Continuous monitoring watches this traffic, looking for unusual patterns. For instance, if a computer in your accounting department suddenly starts sending large amounts of data to a server in another country at 3:00 am, that looks suspicious. Normal accounting work doesn't look like that, nor does it occur at 3:00 in the morning. This kind of anomaly might indicate that someone has compromised that computer and is stealing data.

System Logs

Second, security teams monitor system logs. Every computer, server, and security device generates logs, which are records of what's happening on that device. Someone logged in, a file was accessed, a program was installed, a firewall blocked a connection; these are all examples of events that generate log entries. Individually, these events might seem mundane, but when you collect and analyze them together, patterns may emerge. Continuous monitoring systems collect these logs from across your entire environment and look for patterns that might indicate a security problem.

User Activity

Third, security teams monitor user activity. This means watching what users are doing with their accounts. If someone's account suddenly starts accessing files they've never accessed before, or if they're logging in from unusual locations, these could be signs that their account has been compromised. Remember, attackers often steal legitimate user credentials rather than trying to break through technical defenses, which is why monitoring user activity is so critical.

Individual Devices

Fourth, security teams monitor individual devices: computers, laptops, servers, and mobile devices. What programs are running on them? Are there any suspicious processes? Has any malware been detected? Are there any unauthorized changes to system configurations? This device-level monitoring is where EDR comes in, which we'll discuss in detail shortly.

Vulnerabilities

Finally, security teams continuously scan for vulnerabilities. New security vulnerabilities are discovered constantly. A system that was thought to be secure yesterday might have a newly discovered vulnerability today. Continuous monitoring includes regularly scanning your systems to identify these vulnerabilities so you can patch them before attackers exploit them.

Continuous Monitoring Technologies

Understanding what to monitor leads us to the technologies that make it possible. Modern security operations rely on three interconnected technologies that work together to enable effective continuous monitoring: EDR, SIEM, and SOAR. Don't worry about memorizing the acronyms; what's important is understanding what each tool does.

Endpoint Detection and Response (EDR)

An EDR solution continuously monitors endpoints, such as laptops and servers, for suspicious activity. Think of it as a highly sophisticated security guard stationed on each device, not just checking IDs at the door but constantly watching for suspicious behavior inside.

How does EDR differ from traditional anti-malware software? Remember how we explained that modern anti-malware uses both signature-based detection and behavioral analysis to catch threats? EDR takes those same concepts but operates at an entirely different scale and level of sophistication.

While traditional anti-malware works independently on each device, EDR agents continuously communicate with a central EDR management server. This central server aggregates behavioral data from all endpoints across your organization and uses threat intelligence feeds, which are constantly updated information about emerging threats from around the world, to spot attack patterns that would be invisible to individual devices working alone.

For example, an EDR system might notice that three different workstations in your organization all started exhibiting similar suspicious behavior within minutes of each other. Perhaps all three are trying to access the same unusual server or are attempting to disable security software. A standalone anti-malware program on each computer might not recognize this as part of a coordinated attack, but the EDR's central view connects the dots and raises a high-priority alert.

When a threat is detected, EDR provides powerful response capabilities, such as automatically isolating an infected device from the network to stop an attack from spreading, killing the malicious process, and collecting forensic data for investigation.

Security Information and Event Management (SIEM)

A SIEM is like the central command center for your security operations. It aggregates logs and alerts from all the different security tools across your environment, including your EDR solutions, firewalls, servers, and authentication systems, and collects them in one place for analysis. This "big picture" view is powerful because it allows the SIEM to spot patterns and correlate events that would be invisible if you were only looking at each system individually.

> **EXAMPLE**: Imagine a SIEM sees several events from the same user account in a short period: a failed login attempt on the VPN, followed by a successful login, then access to a file server at 3:00am, and finally a large file transfer to an unknown external IP address. Individually, each event might seem normal on its own, but when the SIEM correlates them, it recognizes a classic attack pattern: compromised credentials being used to steal data. It then generates a high-priority alert for the security team to investigate. The security team can now investigate this incident within minutes instead of potentially missing it entirely.

Security Orchestration, Automation, and Response (SOAR)

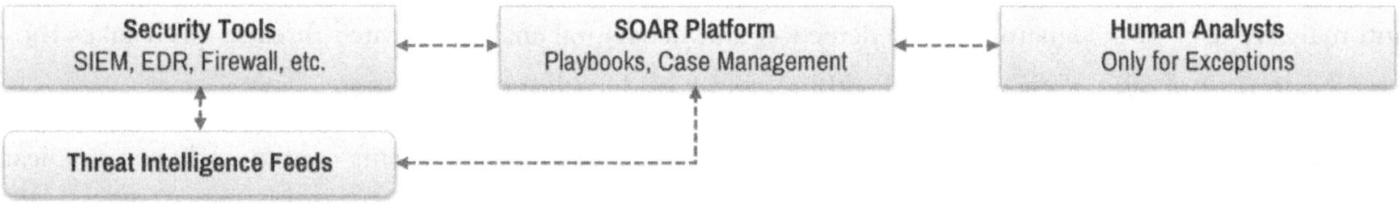

Figure 9.1: SOAR Process

A SOAR platform takes things a step further by automating repetitive security tasks and orchestrating responses across multiple tools. When a SIEM detects a common, well-defined threat, a SOAR platform can automatically execute a pre-written playbook to handle it without waiting for a human analyst to intervene. This automation is important because it helps security teams handle high volumes of alerts more efficiently, allowing analysts to respond faster to routine threats while focusing their attention on complex incidents. For instance, if an EDR tool detects a known piece of malware on a laptop, a SOAR playbook could automatically isolate the device from the network, create a support ticket, and gather forensic data before an analyst even sees the alert. This frees up human analysts to focus on complex investigations.

How They Work Together

These three technologies form an integrated detection and response system. To see how they work as a team, let's walk through a real attack scenario:

- **First, an EDR agent** detects suspicious process behavior on a workstation and sends an alert.
- **The SIEM** receives the EDR alert and correlates it with failed login attempts from the same user on other systems. Recognizing this as a potentially compromised account, the SIEM triggers a high-priority alert.
- **The SOAR platform** instantly executes an automated playbook: it disables the user account, isolates the workstation from the network, and creates an investigation ticket for a security analyst.
- **The analyst** can then investigate safely, using the forensic data collected by the EDR and the aggregated logs in the SIEM to understand the full scope of the attack.

The Human Element and the Security Operations Center (SOC)

While these technologies are powerful, continuous monitoring requires skilled people to use them effectively. Many organizations have a Security Operations Center (SOC), which is a centralized team of security analysts who monitor alerts, investigate potential incidents, and provide 24/7 response capability.

The analysts working in a SOC are like detectives, investigating security alerts and hunting for threats. A SOC is typically organized into several levels, or tiers, each with a specific role:

- **Tier 1 Analysts**: These analysts are the frontline of the SOC. They monitor security dashboards, review incoming alerts, and perform initial triage to determine if an alert is a true threat or a false positive. They handle routine incidents using established procedures and escalate more complex issues to Tier 2.
- **Tier 2 Analysts**: When an incident is escalated, these seasoned incident responders take over. They perform in-depth investigations into complex threats, analyze forensic data, and coordinate the response activities to contain and eradicate the threat.

- **Tier 3 Analysts**: These are the expert-level threat hunters and forensic specialists who handle the most complex and critical incidents. They perform advanced malware analysis, develop new detection methods, and often mentor junior analysts.

Let's see how these three tiers collaborate during a security incident:

- **9:15 AM - Tier 1**: An analyst monitoring the security dashboard notices an alert: A user account is attempting to access a database that it has never accessed before. The analyst checks recent activity and sees the login came from an unfamiliar location. This doesn't look normal, so they escalate it to Tier 2.
- **9:30 AM - Tier 2**: The incident responder digs deeper, examining login history and correlating events in the SIEM. They discover the account was logged in from two different countries within 10 minutes, which is physically impossible. The credentials are clearly compromised. They immediately disable the account and isolate affected systems. During the investigation, they find sophisticated malware on the compromised user's workstation that captured the credentials. This requires expert analysis, so they escalate it to Tier 3.
- **10:00 AM - Tier 3**: The malware specialist begins using advanced forensic tools to reverse-engineer the malware. Over the next several hours, they identify its command-and-control servers and create new detection signatures to deploy across all endpoints. Meanwhile, Tier 2 coordinates with IT operations to force a password reset for the compromised account and verify that multi-factor authentication is enabled before restoring access. The Tier 3 analyst also briefs the team on specific indicators to watch for in future attacks.
- **Result**: This example highlights how the three tiers working together contained the immediate breach within 45 minutes (by disabling the account and isolating systems), prevented data theft, and strengthened the organization's defenses through detailed malware analysis completed later that day.

This human element is essential because security requires judgment, context, and the ability to understand complex situations that can't be fully automated.

Now that you understand the team responsible for day-to-day security (the SOC), let's explore the formal process they follow when a real threat gets through. No matter how good your monitoring is, incidents will eventually occur, and having a structured plan to address them is critical. This structured plan is called incident response, which we'll explore in detail next.

Incident Response

No matter how good your security controls are, incidents will happen sooner or later. A user will click on a phishing link, a vulnerability will be exploited, or a piece of malware will slip through your defenses. When these incidents occur, how you respond makes all the difference between a minor inconvenience and a catastrophic breach.

Incident response is the structured approach to handling security events and breaches. It's about having a plan, following a process, and responding effectively to minimize damage and recover quickly. While a SOC is often central to this effort, incident response is a broader organizational process that may coordinate everyone from technical staff to legal and communications teams.

Not all incidents require an organization-wide response. The SOC is designed to handle routine security events independently, such as blocking common malware or investigating low-level alerts. However, for major incidents, such as a significant data breach, a ransomware attack on critical systems, or any event with legal or public relations implications, the SOC's role shifts. It becomes the technical core of a broader incident response effort that involves legal, communications, HR, and executive leadership.

Remember the Target breach? One of the critical failures was in incident response. Target detected the threat but failed to respond appropriately, allowing the breach to continue for weeks. A well-executed incident response process could have stopped the breach before millions of credit cards were stolen. Let's explore how incident response works so you understand what should happen when things go wrong.

Understanding What an Incident Is

Before we dive into the response process, let's clarify what we mean by a "security incident." Not every security event is an incident. If your firewall blocks a suspicious connection attempt, that's a security event, but it's not necessarily an incident. In fact, it's a good thing because your defenses worked as intended. An incident is a security event that has actually caused harm or has the potential to cause harm. Examples include:

- A computer infected with malware
- A user account that's been compromised
- Unauthorized access to sensitive data
- A successful phishing attack
- A data breach
- A denial-of-service attack that disrupts your services

Some of these incidents, like an infected computer or a compromised user account, are typically handled entirely by the SOC using established procedures. Others, like a data breach or unauthorized access to sensitive data, require coordination with legal, communications, and executive leadership due to their potential business, regulatory, or reputational impact. The key is that something's gone wrong or is going wrong, and you need to take action to address it.

Walking Through the Seven Steps of Incident Response

Figure 9.2: Incident Response Process

Incident response follows a structured process with seven key steps. This process guides you from detecting a security problem to fixing it and preventing it from happening again. Let's walk through each step.

Step 1: Detection

Detection is the process of discovering that an incident has occurred or is occurring. This might happen through security tool alerts, user reports, or external notifications. These alerts are often triggered by the very threats we've discussed in previous chapters, such as a user clicking on a phishing email (Chapter 7) or malware attempting to spread across the network (Chapter 5).

When you receive an alert or report, the first task is triage. This term, borrowed from emergency medicine, involves determining whether this is really an incident and how serious it is. Not every alert is a real problem. A security analyst gathers initial information to decide: Is this real? Is this serious? Do we need to activate our full incident response process? This judgment is crucial because you don't want to treat every minor issue as a major crisis, but you also can't dismiss something serious as a false alarm.

Step 2: Response

Once you've confirmed a real incident is occurring, you mobilize your incident response team. The SOC provides the technical core of this team, but depending on the incident's severity, the response may also include system administrators, managers, legal counsel, and communications staff. Think of it like a hospital emergency room assembling specialists for a serious case.

You conduct an initial assessment to answer key questions: What happened? How did it happen? What systems are affected? What data might be compromised? How serious is this? These answers guide your response strategy. You establish secure communication channels—if attackers might have access to your email, you don't want to discuss response plans there. Teams often use secure messaging apps or phone calls for this reason.

Step 3: Containment

Containment stops the incident from getting worse, like containing a fire before it spreads. The specific actions depend on the incident type. For a malware infection, a common containment step is to isolate the computer from the network to prevent the threat from spreading, while still preserving the device for investigation. For a compromised user account, you'd immediately disable the account or force a password reset to cut off the attacker's access.

The challenge is balancing security with business operations. Shutting down every potentially affected device is safest, but could also halt business. Incident response teams make risk-based decisions about the minimum necessary containment while keeping critical functions running.

Another consideration: Sometimes you don't want to immediately tip off sophisticated attackers. Suddenly shutting down systems alerts them that they've been detected, potentially causing them to accelerate attacks, destroy evidence, or activate backup access. In some cases, quietly monitoring while preparing for complete removal is better, though this requires careful judgment.

Step 4: Reporting

Throughout the incident response process, you need to keep various people informed. Internally, you must update management and executives with appropriate detail levels—executives need high-level summaries while technical teams need detailed information.

Externally, you may have legal obligations to report certain incidents. Many regulations require reporting data breaches within specific timeframes to affected individuals, regulatory agencies, and, in some cases, the media. Requirements vary by jurisdiction and industry. You might also need to report to law enforcement (especially for criminal activity), cyber insurance providers (for potential claims), and customers or partners if the incident affects them. Good documentation is essential for clear reporting about what happened, when, what was affected, and your response.

Step 5: Recovery

Recovery involves two main activities: Eradicating the threat and restoring affected systems. Eradication means completely removing the threat. You must delete malware from all infected systems, ensure attackers no longer have access to compromised accounts, and patch exploited vulnerabilities. The keyword is "completely" because any remaining trace might allow the threat to resurface. This is complex because sophisticated attackers can establish multiple footholds via backup accounts, multiple malware pieces, or several exploited vulnerabilities. For this reason, you must find and remove all of them.

Once you're confident the threat is eradicated, you can begin restoring systems. This involves several critical actions, such as:

- Rebuilding compromised systems from clean system images
- Reapplying security configurations
- Restoring data from clean backups
- Resetting passwords for affected accounts
- And testing systems for proper operation

Restoration must be careful and methodical. Avoid rushing things and accidentally restoring infected systems or reconnecting systems before proper cleaning. You should also monitor restored systems closely using continuous monitoring tools (EDR, SIEM) to verify that the threat is truly gone and that systems are operating normally.

Step 6: Remediation

While recovery focuses on returning to normal, remediation (which is different from recovery) focuses on fixing the underlying problems that allowed the incident to occur in the first place, thereby preventing recurrence. Remediation might involve implementing new security controls, updating policies and procedures, providing additional staff training, improving monitoring capabilities, or fixing misconfigurations. For example, if an attacker exploited an unpatched vulnerability, remediation means not just patching that one system, but ensuring all systems are patched and either implementing or improving the current patch management process to prevent similar exposures. The goal: make the same attack much harder, or even impossible, to repeat.

Step 7: Lessons Learned

The final step is a post-incident review or "lessons learned" session, where the incident response team and stakeholders discuss what happened and how the response went, covering these key questions:

- **Detailed Timeline**: What exactly happened?
- **Root Cause**: How did the attacker get in?
- **Successes**: What did we do well?
- **Improvements**: What could we have done better?
- **Challenges**: Were there surprises or unexpected challenges?
- **Action Items**: What should we change going forward?

The tone matters because this isn't about blaming people; it's about learning and improving. If an employee clicked a phishing link, ask, "Why didn't our email filter catch this?" and "How can we better train employees?" rather than, "Why was this employee careless?" The output should be specific action items: incident response plan updates, new security tools or controls to implement, security policy changes, additional training needs, and process improvements.

Why This Process Matters

You might wonder why we need such a structured process. Can't we just fix problems when they occur? The reality is that incidents are chaotic and stressful. When you're in the middle of a breach with systems down and data potentially being stolen, it's easy to panic and make mistakes. That's where a structured process becomes invaluable. It helps you stay organized and ensures you don't miss critical steps, while also maintaining consistency. Different people might handle different incidents differently, but following the same process yields consistent, effective responses. Finally, the process ensures you learn from incidents. Without reporting and lessons learned steps, you might fix the immediate problem but miss opportunities to prevent future incidents.

> **KEY POINT**: Think back to the Target breach. If Target had followed proper incident response, the outcome would have been very different. Their FireEye system detected the malware and sent multiple alerts (step 1), but the security team failed to investigate or act on those warnings (step 2). As a result, they never got to the crucial containment step, and the breach continued for weeks. A structured incident response process, properly followed, could have stopped the breach early and prevented the theft of millions of credit cards.

Incident response is a crucial reactive capability that helps you respond to attacks as they occur. But what if you could find your weaknesses before an attacker does? An even better approach to security is to be proactive.

Instead of just waiting to respond to threats, how can organizations find their own weaknesses before the attackers do? This is the goal of security assessments.

Security Assessments

Incident response is a crucial reactive capability that helps you respond to attacks as they occur, but identifying your weaknesses before an attacker does is preferable. This is where security assessments come in. While continuous monitoring and incident response focus on detecting and responding to active threats, security assessments take a proactive approach. They involve systematically evaluating your security posture to identify weaknesses before attackers can exploit them. Think of security assessments as preventive medicine for your cybersecurity program. Rather than waiting until you're sick (breached), you conduct regular checkups to identify and address health issues (vulnerabilities) before they become serious problems.

Why We Need Security Assessments

Here's a fundamental truth about cybersecurity: you can't protect what you don't know about, and you can't fix problems you haven't identified. Security assessments help you answer critical questions:

- What vulnerabilities exist in our systems?
- Are our security controls actually working?
- Where are we most at risk?
- Are we meeting our security and compliance requirements?

Without regular assessments, you're essentially flying blind. You might think your security is good, but you won't know for sure until you test it; otherwise, an attacker may find a weakness you didn't know existed.

Types of Security Assessments

There are several different types of security assessments, each serving a different purpose. Let's explore the main ones you should understand.

Vulnerability Assessments

A vulnerability assessment is like a health screening that checks for known problems. It involves scanning your systems (computers, servers, network devices, and applications) to identify known vulnerabilities. Remember the hacking methodology we discussed in Chapter 2? Vulnerability assessments follow the same systematic approach that both black hat and white hat hackers use in the early stages of their operations. Like all hackers, whether malicious or ethical, vulnerability assessments begin with Step 1: Information Gathering and Step 2: Fingerprinting and Scanning to understand what systems exist on your network and what's running on them. They then move into Step 3: Vulnerability Assessment, systematically scanning systems to identify weaknesses. The key difference is that vulnerability assessments stop here at Step 3. Unlike penetration tests (which we'll be discussing next), they don't proceed to Step 4: Exploitation or Step 5: Post-Exploitation. They identify the vulnerabilities but don't attempt to exploit them.

So, how do vulnerability assessments actually find these weaknesses? They use automated tools, such as Nessus or Qualys, to check your systems against databases of known vulnerabilities. The scanner might check things like:

- Are there any missing security patches on your operating systems?
- Are any applications running outdated versions with known security flaws?
- Are there any weak or default passwords?
- Are any unnecessary services running that could be exploited?
- Are there any misconfigurations that create security weaknesses?

The output is a report that lists all the vulnerabilities found, typically with severity ratings. Think of severity ratings as a triage system. They help you prioritize which problems need immediate attention versus which ones can wait. A critical vulnerability might be something like an unpatched server that could allow an attacker to take complete control, whereas a low-severity vulnerability might be an outdated version of software that has a minor security flaw.

The value of vulnerability assessments is that they're relatively quick and easy to run, they can identify many common security problems, and most importantly, they're non-invasive, meaning they don't break things or disrupt operations the way penetration testing sometimes can. The limitation is that they only find known vulnerabilities, meaning problems that are already known and in vulnerability databases. They won't find zero-day vulnerabilities or unique security issues specific to your custom applications. Organizations typically run vulnerability scans frequently, typically weekly or even daily for critical systems. This is important because new vulnerabilities are discovered constantly. A system that was considered secure last week might have a newly discovered vulnerability this week.

Penetration Testing

Penetration testing takes things a step further. Instead of just identifying vulnerabilities, penetration testers actually try to exploit them, simulating what a real attacker would do. A penetration test follows the complete hacking methodology we explored in Chapter 2, moving through all five phases: information gathering, fingerprinting and scanning, vulnerability assessment, exploitation, and post-exploitation. Think of it this way: A vulnerability assessment tells you that your backdoor has a weak lock. A penetration test actually tries to pick that lock, get inside, and see what a burglar could steal once they're in.

Penetration testers use specialized tools and frameworks, such as Metasploit and Burp Suite, to exploit the vulnerabilities they discover. During the exploitation phase, testers try to gain unauthorized access by exploiting unpatched servers, using social engineering tactics like phishing emails, or finding ways to bypass firewalls. If they successfully gain access, they move into post-exploitation, exploring what damage a real attacker could cause:

- Can they access sensitive data?
- Can they move laterally to other systems?
- Can they escalate their privileges to gain administrative control?
- Can they maintain persistent access?

These are questions ethical hackers ask and try to accomplish during the post-exploitation phase, which helps them understand the real-world impact of a successful attack and reveals how far an attacker could penetrate your defenses.

Finally, they document everything they did and provide a detailed report showing what vulnerabilities they exploited, what they were able to access, and recommendations for fixing the problems.

Penetration testing provides valuable insights that vulnerability assessments can't. It shows you not just what vulnerabilities exist, but whether they can actually be exploited and what the consequences would be. It tests your defenses under realistic conditions. However, penetration testing is more time-consuming and expensive than vulnerability scanning. It also carries real risks because penetration tests can break things. Systems might crash, services could be disrupted, or configurations could be altered in ways that cause problems. This is why it's critical to understand these risks before undertaking or agreeing to a penetration test, and why the work requires careful planning, skilled testers, and a clear understanding of what systems are being tested and when. Organizations typically conduct penetration tests less frequently, perhaps annually or after major changes to their environment, for these reasons.

Security Audits

While vulnerability assessments and penetration testing focus on technical security, security audits take a broader view. They examine your entire security program, not just the technology, but also your policies, procedures, and practices. A security audit might review things like:

- Do you have documented security policies, and are they being followed?
- Are you meeting regulatory requirements (like HIPAA for healthcare or PCI DSS for payment cards)?
- How do you manage user access? Are people given appropriate permissions?
- Do you have an incident response plan, and has it been tested?
- How do you train employees on security?
- How do you manage third-party vendors who have access to your systems?
- Are you properly protecting sensitive data?

Audits often involve reviewing documentation, interviewing staff, and examining how security processes actually work in practice. These audits can be conducted internally by your own staff (internal audits) or by independent third-party auditors (external audits), with external audits often required for regulatory compliance. The goal is to ensure that your security program is comprehensive and effective, not just that your technology is configured correctly. Security audits are particularly important for compliance. Many regulations require regular audits to verify that organizations are meeting their security obligations. Even if you're not subject to specific regulations, audits help ensure your security program is well-rounded and addresses all the important areas, not just technology.

Risk Assessments

Risk assessments take the broadest view of all. They're about understanding and prioritizing risks to your organization by evaluating which assets you need to protect, which threats you face, and which vulnerabilities exist. We'll explore risk management in detail in Chapter 10, including how to conduct risk assessments, calculate risk levels, and develop risk treatment strategies. For now, understand that risk assessments help you make informed decisions about where to focus your security efforts.

For example, a risk assessment might identify that your customer database is a high-value asset vulnerable to SQL injection attacks, which would represent a high-priority risk. Meanwhile, an old system with some vulnerabilities but no sensitive data and no Internet access represents a lower-priority risk. Risk assessments help you make informed decisions about where to invest your security resources. You can't fix everything at once, so you need to prioritize based on which risks pose the greatest danger to your organization.

The Essential Bookends: Pre-Engagement and Reporting

Regardless of which type of assessment you're conducting, a vulnerability assessment, penetration test, security audit, or risk assessment, all share two critical phases that bookend the assessment process, just as we saw with ethical hacking in Chapter 2:

- **Pre-Engagement**: This phase precedes any assessment work and establishes the ground rules, scope, objectives, and formal authorization. Just as ethical hackers need legal consent before testing, all security assessments require clear authorization and defined boundaries.
- **Report Writing**: After the assessment work is complete, all findings are documented, analyzed, and presented to stakeholders, including an executive summary, detailed findings with severity ratings, a business impact analysis, and prioritized remediation recommendations. A good report doesn't just identify problems; it provides a clear roadmap for improving security.

Pre-engagement and reporting aren't just administrative formalities; they're essential components that ensure assessments are conducted safely, legally, and productively. Without proper pre-engagement, you risk testing the wrong systems or causing unintended disruption. Without thorough reporting, your findings won't translate into actual security improvements.

Making Assessments Valuable

Conducting security assessments is only valuable if you actually use the results to improve your security. This seems obvious, but it's a surprisingly common pitfall. Organizations spend time and money on assessments, generate reports full of findings, and then, nothing happens. The reports sit on a shelf, and the vulnerabilities remain unfixed. To make assessments valuable, you need to:

- **Prioritize the Findings**: Not everything can be fixed immediately. Focus first on the most critical issues: the ones that pose the greatest risk.
- **Create Action Plans**: For each significant finding, develop a specific plan for addressing it. Who'll fix it? When will it be fixed? What resources are needed?
- **Track Progress**: Monitor remediation efforts to ensure issues are actually being addressed. Don't let things fall through the cracks.
- **Verify Fixes**: After vulnerabilities are remediated, verify that the fixes actually worked. Re-scan or re-test to confirm the problems are resolved and that the fixes didn't introduce new vulnerabilities.
- **Report to Management**: Keep leadership informed about assessment findings and remediation progress. Security assessments often reveal issues that require management attention or resources.

The goal is to create a continuous improvement cycle: assess, identify problems, fix them, and then assess again to verify improvement and find new issues.

Bringing It All Together

Security operations is where cybersecurity theory meets practice. While previous chapters focused on understanding threats and implementing defensive controls, this chapter has shown you how organizations maintain security day to day through continuous vigilance and a structured response.

The core components work together as an integrated system. Continuous monitoring through EDR, SIEM, and SOAR provides real-time visibility across your environment. Incident response provides a structured seven-step process for effectively handling security events. Security assessments proactively identify vulnerabilities before attackers can exploit them. They move through the same hacking methodology we learned in Chapter 2, but stop at different phases depending on the assessment type.

The Target breach illustrates the critical importance of effective security operations. Target had sophisticated detection tools that identified the malware and alerted their security team, but their security team failed to investigate and respond. This breakdown demonstrates that security operations require more than just technology; it demands the right processes, skilled people, and an organizational culture that acts decisively on threats.

The key lesson is that security is not set-it-and-forget-it. Detection without response is meaningless. Tools without proper processes fail. And processes without skilled people to execute them accomplish nothing. Effective security operations integrate all three elements (technology, process, and people) into a continuous cycle of monitoring, detection, response, assessment, and improvement.

As threats evolve, your security operations must evolve with them. Organizations that learn from incidents, continuously assess their security posture, and act on their findings become more resilient over time. This ongoing practice of vigilance and improvement is what separates organizations that prevent or minimize breaches from those that suffer catastrophic incidents.

Key Takeaways

- **Security operations requires continuous vigilance** through ongoing monitoring, detection, and response, not one-time implementations.
- **Detection without response is meaningless.** The Target breach showed that alerts are worthless unless acted on promptly.
- **EDR, SIEM, and SOAR work together** to provide visibility, correlation, and automated responses across your environment.
- **A structured incident response process is essential.** The seven-step process ensures consistent, effective handling of security events from detection through lessons learned.
- **Security assessments follow the hacking methodology.** Vulnerability assessments stop at Step 3, while penetration tests complete all five phases.
- **Proactive assessments prevent breaches** by identifying and fixing problems through regular vulnerability scans and penetration tests before attackers can exploit them.
- **Pre-engagement and reporting are essential.** These bookend phases ensure assessments are conducted safely, legally, and productively.
- **Effective security operations require integrating** people, process, and technology. All three elements must work together for effective defense.

Quick Action: Start Your Security Operations Journey

In this chapter, you learned about security operations: how organizations actively defend against threats through monitoring, incident response, and proactive assessments. Now put these concepts into practice with two tasks: checking whether your accounts have been compromised and, optionally, running a vulnerability scan of your home network.

What to Do

Check for Breached Accounts:

- Security operations teams constantly monitor for compromised credentials, and you can do the same for your personal accounts.
- Visit Have I Been Pwned (haveibeenpwned.com) and enter your email address to see if it appears in any known data breaches.
- If any accounts show up, investigate which services were compromised and update your login credentials for those accounts, along with any other accounts where you reused the same credentials.

Optional: Run a Vulnerability Scan on Your Home Network:

- Professional security teams use vulnerability scanners like OpenVAS and Nessus to identify weaknesses before attackers do, and you can run these tools at home, too.
- Some great YouTube tutorials walk you through setting up OpenVAS and Nessus. Search YouTube for "OpenVAS tutorial" or "Nessus Essentials home lab" to find step-by-step setup guides.
- Try running a scan to see which vulnerabilities might exist on your home network and devices. You'll gain hands-on experience with the same tools that security operations teams use daily.
- **Note**: Setting up these tools requires some technical comfort. Don't be discouraged if this feels challenging. Security professionals spend time learning these tools, too.
- **Important**: Only scan devices and networks you own on your own home network. Scanning without permission is illegal.

By completing this breach check, you've performed a basic security assessment of your digital footprint. If you've tackled the vulnerability scan, you've gained hands-on experience with the same tools professional security teams use to find vulnerabilities before attackers do.

Chapter 10. Security Management & Culture

Introduction

The previous chapter covered security operations, the day-to-day practice of monitoring, detecting, and responding to security threats. We learned about the tools and processes that keep organizations safe on an operational level. But security operations don't exist in a vacuum. For it to be effective, it needs to be supported by strong management practices and a security-conscious culture throughout the organization.

Think of it this way: Security operations is like having a well-trained fire department with excellent equipment and procedures. But what if buildings aren't constructed with fire safety in mind? What if people don't know how to prevent fires or what to do when one starts? The fire department can only do so much if the broader environment doesn't support fire safety.

This is where security management and culture come in. Security management encompasses the policies, processes, and programs that create a framework for security across the organization. Security culture is about creating an environment where everyone, not just the security team, understands their role in keeping the organization safe.

In this chapter, we'll explore these key components. We'll start with the legal and regulatory landscape that shapes security programs, then look at how organizations formalize their approach to security through governance and risk management. Finally, we'll examine the human side of security—the personnel policies that set expectations for behavior, and the awareness and training programs that help people understand their security responsibilities. Throughout, we'll see how these elements work together to create an environment where security is woven into the fabric of how the organization operates.

Legal & Regulatory Compliance

Before we dive into the operational aspects of security management, we need to understand a fundamental point: the legal and regulatory requirements that drive much of what organizations do in cybersecurity. You might be wondering why we're starting with laws and regulations. Here's why: compliance isn't optional, and understanding these requirements helps you appreciate why organizations invest so heavily in security programs.

Organizations don't implement security programs just because it's a good idea, though it certainly is. They often do so because laws and regulations require it. Depending on your industry, you might be subject to certain federal laws, state regulations, or industry-specific requirements. The specific requirements vary by industry, but one principle applies universally: ignorance of the law isn't a valid excuse.

If you're driving 45 mph in a 30 mph zone and you tell the police officer, "I thought the speed limit was 45," that won't get you out of the ticket. The law still applies whether you knew about it or not. The same principle applies to cybersecurity compliance: not knowing about a regulation doesn't exempt you from following it.

If you're not in compliance, you can be held legally liable. This liability could arise under:

- **Criminal Law**: for intentional violations, which could result in jail time
- **Civil Law**: for damages caused by non-compliance, where you might be sued by affected parties
- **Administrative Law**: for regulatory violations, where government agencies can impose fines and sanctions

The consequences can include fines, lawsuits, criminal charges, and severe damage to your organization's reputation.

Common Compliance Requirements

While you don't need to be a legal expert, it's important to be aware of the major regulations that shape cybersecurity practices. Here are some of the most common ones you're likely to encounter:

- **Federal Information Security Management Act (FISMA)** applies to U.S. federal government agencies and contractors. It requires them to implement comprehensive security programs and to report security incidents to authorities such as the U.S. Computer Emergency Readiness Team (US-CERT).
- **Health Insurance Portability & Accountability Act (HIPAA)** applies to healthcare providers, such as doctors and hospitals, and anyone who handles patient health information. It mandates strict protections for the confidentiality, integrity, and availability of health records.
- **Gramm–Leach–Bliley Act (GLBA)** applies to financial institutions such as banks, credit unions, and insurance companies. It requires them to develop written security plans and implement safeguards to protect customer financial data.

- **Payment Card Industry Data Security Standard (PCI DSS)** is an industry standard, not a law, but it's mandatory for any business that accepts credit card payments. If you want to process transactions from Visa, Mastercard, or American Express, you must comply with their security requirements.
- **General Data Protection Regulation (GDPR)** is a European Union law that has a global impact. It sets strict rules on how organizations worldwide must handle the personal data of EU citizens.

While these are just a few examples, the message is clear: You need to identify which laws, regulations, and standards apply to your organization or industry. Once you've identified them, you also need to ensure that your security program addresses their requirements.

Due Care, Due Diligence, & Gross Negligence

Now, we'll discuss three legal concepts that frequently appear in cybersecurity and help determine whether an organization has acted responsibly. It's important for me to mention that I'm a cybersecurity professional, not an attorney, so we'll be looking at these from a practical security perspective. Understanding them is crucial, as they directly impact how security programs are assessed.

Due care means doing what any reasonable, responsible person would do in a given situation. It's often called the "prudent person rule." In cybersecurity, this means implementing appropriate security measures to protect against known risks. For instance, if you're a systems administrator and there's a known vulnerability with an available patch, due care means testing and installing that patch in a timely manner. That's what a prudent person would do.

Due diligence is the management of due care. It means ensuring, over time, that due care is actually being performed. While due care is about doing the work, due diligence is about verifying that the work is being done correctly and consistently. For example, regarding patch management, management exercises due diligence by reviewing patch management reports to confirm systems are up to date, and following up when vulnerabilities aren't being addressed. This same principle applies to all security activities. Management must verify that security work is being done correctly and consistently.

Gross negligence is the opposite of due care and due diligence. It means failing to exercise even the most basic level of care that a reasonable person would exercise. In other words, it represents failing to meet even basic reasonable standards, showing a reckless disregard for security. If an organization isn't performing due care and management isn't exercising due diligence, and this failure leads to harm or loss, the organization may be held legally liable for gross negligence.

Remember the Equifax breach from Chapter 1? While the legal settlement didn't explicitly label this as gross negligence, security and legal experts widely view Equifax's failure to patch a known vulnerability for months as a clear example of how failing to exercise due care and due diligence can result in massive legal and financial consequences. The result was over $700 million in settlements and significant reputational damage.

> **KEY POINT**: Understanding these legal concepts helps you appreciate why security isn't optional. Organizations that exercise due care and due diligence meet their legal obligations and reduce their risk. Those that don't may face not only security breaches but also legal liability for failing to act responsibly.

Information Security Governance

With the legal landscape defined, we can explore how organizations actually govern their security efforts. You might be thinking: "Isn't this just management?" Not quite. Governance is specifically about strategic direction, oversight, and accountability, specifically, the "who decides what" and "how decisions get made" of cybersecurity.

Information security governance is the process by which an organization directs and controls its information security program through policies, procedures, and clearly defined roles and responsibilities. Think of it as the organizational blueprint for security. It defines who has authority to make decisions, how those decisions are made, and how the security program is coordinated across the entire organization.

Organizations typically have governance program management offices (PMOs) or personnel specifically assigned to governance roles. In smaller organizations, governance responsibilities might be handled by the CISO or security director, but the principles remain the same. These individuals work with senior leadership to create the structure, policies, and processes that guide security activities. Governance provides strategic direction and ensures that cybersecurity objectives, such as effective risk management, are achieved.

Governance answers critical questions that every organization faces:

- How much security is enough for our organization?
- Which security investments should we prioritize?
- How do we balance security with business needs?
- Who has the authority to make security decisions?
- How do we ensure security supports rather than hinders business objectives?

Determining "How Much Security Is Enough"

One of governance's most important roles is answering a deceptively simple question: how much security is enough? This isn't a technical question; it's a business question that requires balancing multiple factors.

Imagine an organization is securing its network. Should it deploy a basic firewall or an enterprise-grade next-generation firewall? Does it need an intrusion detection system or an intrusion prevention system? Should it implement a honeypot? The answer isn't "implement the maximum security possible" because security has costs, including direct costs (purchasing tools), operational costs (staff time), and productivity costs (controls that slow down work).

Governance helps by establishing processes for making informed decisions. Organizations evaluate security investments by considering:

- **Compliance Requirements**: Sometimes, regulations mandate specific controls. For example, FISMA or PCI DSS might require certain measures regardless of their return on investment (ROI).
- **Risk Reduction Value**: For security decisions not driven by compliance, you want controls that deliver a positive ROI, where the value of risk reduction exceeds the cost of implementation.
- **Business Impact**: Governance ensures security decisions consider business needs. Remember, a technically sound security control that cripples productivity isn't always the right choice.

These governance decisions are typically made by a combination of executive leadership, the CISO or security director, and other key stakeholders, such as legal, compliance, and business unit leaders. Ultimately, governance helps organizations find the appropriate amount of security, not too little, not too much, but the right amount to manage risks to acceptable levels while supporting business operations.

The Essential Role of Senior Management

Information security governance requires active participation from senior management—the executives who lead the organization, such as the Chief Executive Officer (CEO), Chief Information Officer (CIO), and Chief Information Security Officer (CISO).

Why is senior management so crucial to governance? Three key reasons:

1. **They Craft & Approve High-Level Policies**: Senior executives are responsible for creating and signing off on the strategic policies that guide the entire security program.

2. **They Drive Security Dirction Based on Business Needs**: Remember, security exists to support the business. Senior management ensures that security activities align with the organization's strategic goals.
3. **They Make Final Decisions & Allocate Resources**: Major security decisions, especially those involving significant funding, require executive approval.

KEY POINT: Effective security management must be top-down, not bottom-up. When direction comes from senior leadership, everyone below understands they're expected to comply. Without executive support and funding, security programs will struggle to succeed.

Patch & Change Management

Governance provides the strategic direction. Next, we turn to two fundamental processes that keep IT environments secure and stable: patch management and change management.

Patch Management: Keeping Systems Up to Date

Software isn't perfect. Developers write millions of lines of code, and inevitably, some of that code will contain security vulnerabilities that attackers can exploit. As we discussed in Chapter 6, some of these are zero-day vulnerabilities that vendors don't even know about until they're discovered by researchers, often through bug bounty programs. When vendors become aware of these flaws, they release patches (also called updates and hotfixes) to fix the problems.

Patch management is the end-to-end process of testing, deploying, and verifying that patches are successfully applied across your IT environment. In a business with hundreds of computers, it's more complex than clicking "update now."

Here's why patch management requires a structured process:

- **Patches Can Sometimes Break Things**: A patch that fixes one problem might create another. Therefore, you need to test patches before deploying them.
- **Timing Matters**: You can't just shut down critical business systems in the middle of the workday. You need to schedule updates during maintenance windows, typically in the evening or on weekends.
- **You need to verify success**: You should regularly audit your systems to confirm that patches were applied correctly.

The Patch Management Process

Here's how organizations typically manage patches:

- **Testing comes first.** When a vendor releases a new patch, IT teams don't immediately push it to all systems. Instead, they install it on test systems that mirror their production environment to make sure it doesn't cause problems. This testing phase catches issues before they affect users.
- **Deployment happens during maintenance windows.** Once testing confirms the patch works properly, IT schedules its installation on production systems—the live, operational systems people actually use. This typically happens during off-hours, such as evenings or weekends, to minimize disruption. For large environments with hundreds or thousands of systems, patch management tools automate much of this work.
- **Auditing verifies success.** After deployment, IT teams audit their systems to confirm that patches were installed correctly and to identify any systems where installation failed. This verification step ensures nothing slipped through the cracks.

Patch management is one of the most important security practices because unpatched vulnerabilities are among the most common ways attackers break into systems. There's often a race between organizations patching their systems and attackers exploiting them. The longer you wait, the more time attackers have.

> ! **COMMON MISTAKE**: Skipping the testing phase to deploy patches faster. While speed is important for critical vulnerabilities, deploying untested patches can crash systems, break critical applications, or cause unexpected failures. In some cases, the patch causes bigger problems than the vulnerability itself.

Change Management: Controlling IT Changes

While patch management focuses on updates, change management addresses a broader question: how do organizations ensure that any change to their IT environment doesn't introduce new problems or security risks?

In any organization, the IT environment is constantly evolving. New servers are deployed, applications are upgraded, and network configurations are modified. Each of these changes has the potential to cause problems. Change management is the governance mechanism that ensures major changes don't occur without proper review and approval. Think of change management like a building permit process. Before you make any significant changes to your house, you need a permit to ensure your plans meet building codes and won't create any safety hazards. Change management serves a similar purpose for IT.

How Change Management Works

Let's walk through a typical scenario. Imagine a team wants to deploy a new database server. Before they can move it into production, they prepare a change request that documents what they're doing, why, the risks, and their rollback plan. This request is presented to a change control board (sometimes called a change advisory board), a group of stakeholders from IT operations, security, business units, and management. They review the request and ask questions, such as:

- Have the security requirements been addressed?
- What's the impact if something goes wrong?
- Is there a rollback plan if the change needs to be reversed?
- Has the change been properly tested?

If the board is satisfied, they'll approve the change. If not, they might reject it or ask for more work or additional information.

Why Change Management Matters

- It **prevents mistakes** by forcing thorough planning and review.
- It **ensures security requirements are met** by making security an explicit part of the approval process.
- It **maintains stability** by preventing uncontrolled changes, a major source of IT outages.
- It **provides accountability** by maintaining a clear record of who approved what and why.

Additionally, most organizations have an expedited emergency change process for critical security patches or urgent fixes, but even these typically require after-the-fact review and documentation.

> ! **COMMON MISTAKE**: Treating change management as unnecessary bureaucracy and bypassing the process to "move faster." Uncoordinated changes are a leading cause of outages and security incidents.

Risk Management

In Chapter 4, we introduced the core concepts of risk management. Now, we're going to build on that foundation to understand how organizations use these concepts in a structured way. If you recall, we established that cybersecurity is fundamentally about managing risk. Let's now explore how organizations build a systematic approach around these ideas.

Understanding Risk

In Chapter 4, we introduced risk as the intersection of a threat and a vulnerability. Let's dive deeper into that idea. Risk has three key elements that must all be present:

- **Threat**: Something or someone that could cause harm, such as hackers, malware, natural disasters, or human error.
- **Vulnerability**: A weakness that could be exploited, such as unpatched software, weak passwords, or missing backups.
- **Impact**: The potential damage if a threat exploits a vulnerability, such as a data breach, system outage, or financial loss.

Risk exists only when all three elements are present. This leads us to a simple way of thinking about risk:

Risk Equation: Risk = Threat x Vulnerability

Figure 10.1: Risk Equation

When either threats or vulnerabilities increase, risk increases. Security controls work by reducing one of these factors.

Consider a house with a broken lock (vulnerability). If you live in an area with no crime (no threat), the risk is low. But suppose you live in a high-crime area (threat present) and keep valuables inside (impact). In that case, that same broken lock becomes a serious risk. The vulnerability hasn't changed, but the threat level and potential impact are much higher.

Why Risk Management Matters

Perfect security is impossible, and every organization has limited resources in terms of budget, staff, and time. You can't protect against everything, so how do you decide where to focus your efforts? This is where risk management comes in. It's the process of identifying, assessing, and prioritizing risks so you can make informed decisions about how to handle them.

Risk management isn't about eliminating all risk—that's impossible. It's about making smart decisions to reduce risk to an acceptable level and focusing your limited security resources where they'll have the greatest impact.

Risk Assessment

Risk assessment is the process of identifying and evaluating risks. While large organizations use complex quantitative methods, most start with qualitative risk assessment, which consists of evaluating risks based on informed judgment.

Here's how a basic qualitative risk assessment works. For each risk you identify, you ask two questions and assign a score:

- **Likelihood**: How likely is this to happen?
- **Impact**: If it happens, how bad will it be?

You then assign each a score, typically 1 for Low, 2 for Moderate, and 3 for High, and multiply the scores to get an overall risk rating. This helps you prioritize which risks deserve immediate attention. This is often visualized with a risk matrix.

Impact			
High (3)	3	6	9
Moderate (2)	2	4	6
Low (1)	1	2	3
	Low (1)	Moderate (2)	High (3)

Likelihood

Figure 10.2: Risk Matrix

Let's see how this works with an example. Imagine you're the IT manager for a small medical practice, and you've identified two potential risks:

Risk #1: Ransomware Attack

The practice's staff opens patient and insurance company emails every day, and ransomware campaigns are specifically targeting healthcare organizations. A successful ransomware attack would encrypt all the practice's systems, making patient records completely inaccessible. You wouldn't be able to see appointments, review medical histories, or access any clinical data. Patient care would halt immediately, and you'd face difficult decisions about paying the ransom versus rebuilding everything from backups.

- **Likelihood**: High (3) – According to recent reports, healthcare organizations experience ransomware attacks weekly, and the practice lacks advanced email filtering and employee security training.
- **Impact**: High (3) – Complete operational shutdown, potential HIPAA violations requiring notification to patients and regulators, ransom payments potentially reaching hundreds of thousands of dollars, and weeks of recovery time.
- **Risk Score**: 3 × 3 = 9 (Critical)

Risk #2: Medical Practice Affected By Floods

The medical practice is located in a designated flood zone near a river. While flooding doesn't happen every year, major floods have occurred twice in the past decade. A significant flood could damage the entire facility, including the lobby, offices, patient rooms, and IT equipment.

- **Likelihood**: Low (1) – Floods occur in the area every 5-10 years, and the building has flood barriers installed.
- **Impact**: High (3) – The entire practice would need to close for repairs and cleanup, potentially for several weeks. Patient appointments would be cancelled, revenue would stop, and you'd face significant repair costs.
- **Risk Score**: 1 × 3 = 3 (Low)

Both scenarios involve potential damage to the medical practice, but the ransomware attack deserves far more immediate attention because it's far more likely to happen. This is how risk assessment helps you make rational decisions about where to invest your limited security budget.

Risk Response

As we covered in Chapter 4, once you've assessed your risks, you have four main ways to respond. Here's a quick recap:

- **Avoidance**: Eliminate the risk by not doing the risky activity (e.g., not allowing employees to use personal USB drives on work computers).
- **Mitigation**: Reduce risk by implementing security controls (e.g., installing firewalls and anti-malware software).
- **Transference**: Shift the financial impact of the risk to a third party (e.g., purchasing cyber insurance).
- **Acceptance**: Acknowledge the risk but take no action, usually because the risk is low or the cost to fix it is too high.

It's important to understand that after you apply controls, some **residual risk**, the risk that remains, will almost always be left over. The goal is to reduce risk to an acceptable level, not to zero.

IT Risk Management Program

Now that you understand the fundamental concepts, let's look at how organizations build a formal program around them. A formal IT risk management program creates a structured, consistent approach to identifying, assessing, and managing risks across the entire organization. It ensures risk management happens proactively, not just reactively when problems arise.

The Risk Management Plan (RMP)

The foundation of a formal program is the Risk Management Plan (RMP). Think of this as the blueprint for how your organization will handle risk. The RMP is a document that answers key questions:

- **Process**: What's our process for identifying, assessing, and responding to risks?
- **Prioritization**: What criteria do we use to score risks and decide which ones need immediate attention?
- **Authority**: Who has the authority to make decisions about risk responses?
- **Tracking**: How do we document risks and track the status of mitigation efforts?
- **Escalation**: When should a risk be escalated to management?

Perhaps most importantly, the RMP signals management buy-in. When leadership formally adopts an RMP, they're committing to risk management as a priority. This top-down support is crucial for success because, without it, teams may view risk assessments as obstacles that slow them down rather than as essential security practices.

The Risk Program Management Office (PMO)

While the RMP provides the blueprint, someone needs to implement and manage the program. This is the role of the Risk PMO, a business unit charged with ensuring consistent risk management practices.

The risk PMO serves as the central hub for risk management across the organization. They design the frameworks, templates, and tools that everyone uses, ensuring that the marketing department assesses risks the same way the IT department does. They build risk awareness by educating teams through training and support, and they often lead or coordinate risk assessment activities across different departments. The PMO maintains the organization's risk register, a comprehensive list of all identified risks, and provides regular reports to leadership on the organization's overall risk posture. In essence, they turn abstract risk management concepts into concrete, repeatable processes that the entire organization can follow.

A formal risk management program puts these concepts into practice. To see how this works in the real world, let's look at a case study of a company whose success was built on its ability to manage massive technological risks.

Case Study: Netflix's Calculated Risk for Cloud Success

Not all risk management stories are about preventing disaster. Sometimes, they're about enabling innovation. The story of Netflix's move to the cloud is a powerful example of how a company can embrace significant technology risks to achieve even greater rewards.

Background

In the late 2000s, Netflix was growing at an explosive rate, but its reliance on its own data centers was holding it back. A major database corruption in 2008—which halted DVD shipments for 3 days and cost the company millions in lost revenue and customer trust—was the final straw. The company realized that building and managing its own data centers at a global scale was not only expensive and slow but also a huge risk to its availability. To achieve its ambition of becoming a global streaming giant, Netflix made a bold, calculated decision to migrate its entire infrastructure to a then-unproven public cloud, Amazon Web Services (AWS).

The Challenge & Decision

This isn't a story about fending off an attack; instead, it's about proactively managing immense business and technical risks. At the time, moving a mission-critical, large-scale service to the public cloud was almost unheard of. The leadership team had to weigh two sets of risks:

Risks of Staying On-Premise

- **Limited Scalability**: Building new data centers couldn't keep pace with subscriber growth.
- **High Costs**: Massive capital investment in hardware that might sit idle during off-peak hours.
- **Slow Innovation**: Deploying new features was a slow, manual process.
- **Single Points of Failure**: A single data center outage could bring down the entire service.

Risks of Moving to the Cloud

- **Security Concerns**: Entrusting sensitive customer data to a third-party provider was a major concern.
- **Reliability Fears**: Cloud platforms were new and perceived as less reliable than private data centers, and an AWS outage could cripple Netflix's entire service.
- **Vendor Lock-in**: Becoming completely dependent on a single provider (AWS) created a strategic risk.

Netflix's leadership, exercising a high-risk appetite, decided that the risks of standing still were far greater than the risks of moving forward. The opportunity for elastic scalability, global reach, and faster innovation outweighed the security and reliability concerns of the cloud. They accepted the risk but committed to aggressively mitigating it.

Mitigation & Impact

Netflix didn't just move to the cloud; they re-architected their entire platform with the assumption that "failures will happen." This mindset led to groundbreaking risk mitigation strategies:

- **Designing for Failure**: Instead of trying to prevent every failure, they built systems that could survive them. They spread their infrastructure across multiple AWS regions and availability zones, so if one region or zone went down, traffic would automatically reroute to others. This is a perfect example of implementing geographic dispersal and redundancy, which we'll discuss in the next chapter.
- **Proactive Testing**: Netflix pioneered Chaos Engineering (the practice of intentionally and randomly shutting down parts of its live production systems to test resilience). This proved brilliant because it forced engineers to build services that could withstand unexpected outages and fix weaknesses before a real failure occurred. Netflix even created and open-sourced tools like Chaos Monkey to automate this testing, which other companies now use for their own resilience testing.
- **Aggressive Security Investment**: They invested heavily in cloud security, implementing robust encryption for data at rest and in motion, strict identity and access management controls, and sophisticated monitoring to detect threats in their new cloud environment.

The impact was transformative. Netflix was able to scale globally, handle massive traffic spikes for hit shows like *Stranger Things*, and innovate at a speed its competitors couldn't match. By 2016, they had completed their migration and shut down their last data center. As of 2024, Netflix serves over 270 million subscribers worldwide, a scale that would have been impossible without its risk-based decision to embrace the cloud.

Lessons Learned

- **Risk Management Enables Innovation**: Netflix showed that risk management isn't just about preventing bad things; it's also about enabling good outcomes like innovation and growth.
- **Accepting Risk is a Strategic Decision**: Some risks are worth taking if the potential rewards are high enough. The key is to make a conscious, informed decision and have a robust plan to mitigate the downsides.
- **Build for Resilience, Not Perfection**: Instead of trying to prevent 100% of failures, build systems that can gracefully handle them when they inevitably occur.
- **Proactive Mitigation is Better Than Reactive Recovery**: By intentionally testing their own systems for weaknesses, Netflix found and fixed problems before they could cause a real outage, turning a potential vulnerability into a strength.

Personnel Policies

Technology is only part of security. People are equally important, and in many ways, more challenging. You can configure a firewall perfectly, but if an employee clicks on a phishing link, your defenses are bypassed. This is why personnel policies are a critical component of security management. These policies set expectations for behavior, establish accountability, and help create a security-conscious culture. Let's look at some of the most important ones.

Acceptable Use Policy (AUP)

The Acceptable Use Policy (AUP) is one of the most fundamental security policies and is typically something all new employees must read and sign. Its purpose is to define what employees can and can't do when using company IT resources, such as computers, networks, and Internet access. It sets clear boundaries to prevent actions that could introduce security risks or legal liabilities. A typical AUP might specify:

- **What's Prohibited**: Using company computers for personal business, visiting inappropriate websites, installing unauthorized software, and sharing account credentials.
- **What's Required**: Keeping passwords confidential, reporting security incidents, and using company resources only for legitimate business purposes.
- **What the Consequences Are**: Violations might result in warnings, loss of access privileges, or even termination.

Code of Ethics

While an AUP focuses on the acceptable use of technology, a Code of Ethics addresses broader principles of professional and ethical conduct. It's especially important for people in positions of trust who have access to sensitive information or systems. A Code of Ethics establishes a moral compass for employees, guiding their decisions and actions beyond just technical rules. A Code of Ethics might cover:

- **Ethical Use of Information**: Respecting the confidentiality of sensitive data and not using insider information for personal gain.
- **Ethical Conduct in Security Work**: For security professionals, this means using security tools and access only for authorized purposes and reporting vulnerabilities through proper channels.

Separation of Duties

Separation of Duties is a key control principle designed to prevent fraud and errors by ensuring that no single individual has control over all aspects of a critical transaction. It works by dividing a sensitive process among multiple people, which creates a system of checks and balances.

In financial transactions, for example, the person who creates an invoice and the person who approves the payment shouldn't be the same person. This separation prevents someone from creating a fraudulent invoice and paying it to themselves. Additionally, in software development, the developer who writes code shouldn't be the same person who approves and deploys that code to production systems. This separation ensures that code is reviewed for quality and security issues before it goes live, and prevents a malicious developer from sneaking harmful code into production. Separation of Duties serves two important purposes:

- It **reduces insider threat risk** because an individual can't complete a harmful action alone.
- It also **reduces continuity risk** because if one person is unavailable, others can still perform critical functions.

Remote Work Policy

With the rise of remote and hybrid work, a Remote Work Policy has become essential. This policy establishes guidelines for employees working outside the office, focusing on securing company information in less-controlled environments. It recognizes that security risks differ when employees work from home, in a coffee shop, or in a hotel.

A remote work policy typically mandates:

- **A Secure Work Environment**: Employees are required to work in a private location where confidential information (on screen or on paper) can't be viewed by unauthorized individuals.
- **Secure Network Connections**: Employees must use a company-provided VPN to connect to the internal network, ensuring all traffic is encrypted, and the use of untrusted public Wi-Fi for sensitive work may be prohibited.
- **Protection of Company Equipment**: Employees are responsible for the physical security of company-issued laptops and devices.
- **Incident Reporting**: Employees must immediately report any lost or stolen devices or any suspected security incidents to the IT department.

These are just a few examples of essential personnel policies. The specific policies your organization needs will depend on your industry, size, and risk profile.

> ! **COMMON MISTAKE**: Creating policies that sit on a shelf and are never enforced. Effective policies must be clearly communicated to employees, consistently enforced when violated, and reviewed regularly. Most importantly, they must be reasonable enough that people will actually follow them.

Awareness & Training

Policies tell people what they should do, but they don't always explain why or how. This is where awareness and training come in. They're essential for turning policies into practice and building a strong security culture.

Why Awareness & Training Matter

So why is security awareness and training so important? Because in almost every security incident, human behavior plays a role. Technology alone can't stop a curious employee from clicking a convincing phishing link, an overworked staff member from reusing a simple password, or a well-intentioned person from holding the door for a social engineer. These aren't technical failures; they're human ones. Awareness and training are our primary defenses against these human vulnerabilities.

Security is complex, threats are constantly evolving, and people aren't naturally equipped to recognize and respond to cyberthreats. Awareness and training help bridge this gap by equipping your people with the knowledge to serve as a powerful layer in your defense-in-depth strategy.

Awareness vs. Training

It's helpful to distinguish between awareness and training:

Security awareness is about making people conscious of security issues. It's broad and applies to everyone. Awareness programs might include posters, regular emails about current threats, brief videos, and security tips in company newsletters. The goal is to keep security top of mind.

Security training is more in-depth and focused on building specific knowledge and skills, often for specific roles. For example:

- **All employees** might receive annual training on phishing, password security, and data handling.
- **Developers** might receive training on secure coding practices.
- **IT staff** might receive technical training on specific security tools.

The goal of training is to build real competence, not just awareness.

What Should Training Cover?

For general employee training, topics typically include:

- **Phishing & Social Engineering**: How to recognize suspicious emails and what to do
- **Password Security**: How to create strong passwords and use password managers
- **Data Handling**: How to identify and protect sensitive data
- **Physical Security**: Not leaving computers unlocked or letting strangers into secure areas
- **Incident Reporting**: How and when to report potential security issues
- **Remote Work Security**: Using VPNs and securing home networks

Building a Security-Conscious Culture

Ultimately, the goal of all these policies and training programs is to build a security-conscious culture. This is an environment where security isn't just a set of rules enforced by the IT department; it's a shared value that's embedded in how people work every day.

In a strong security culture, people don't just follow the rules because they have to; they do so because they understand why it's important. They see security as a collective responsibility that protects the company, its customers, and their own jobs. Employees in this environment naturally think about security in their daily tasks and feel comfortable reporting potential mistakes or suspicious activity without fear of blame. They view security not as an obstacle, but as a necessary part of doing business well. This cultural shift transforms your workforce from your biggest vulnerability into your greatest security asset.

You can have the best technical security controls in the world, but if your employees routinely click on phishing links and use weak passwords, your security will be weak. Conversely, if your employees are vigilant, they become a powerful layer of defense that complements your technical controls.

> ! **COMMON MISTAKE**: Treating security training as a one-time annual event that employees click through without paying attention. Effective training is ongoing, engaging, and reinforced through regular awareness activities and real-world simulations like phishing tests.

Bringing It All Together

Security management and culture represent the organizational foundation that enables cybersecurity to succeed. While technology provides the tools and security operations provides the day-to-day vigilance, security management creates the structure, processes, and human awareness that make security sustainable.

The management components work together as an integrated system. Legal and compliance requirements set the mandatory baseline for security. Information security governance provides the top-down strategic direction and accountability from leadership. This direction is then put into practice through structured processes like patch management, which keeps systems updated, and change management, which ensures new deployments are safe and stable.

A formal IT risk management program provides a systematic approach for making informed decisions, helping you focus limited resources on the threats that matter most. We saw in the Netflix case study how taking a mature, risk-based approach to cloud migration enabled bold innovation and competitive advantage—not by eliminating risk, but by managing it strategically.

But processes alone aren't enough—security ultimately depends on people. Personnel policies, such as the AUP and remote work policies, set clear expectations for behavior, while awareness and training equip people with the knowledge and skills to meet those expectations. When leadership visibly supports these efforts, it fosters a security-conscious culture where everyone understands they have a role to play. This integration of governance, risk, and culture is what separates organizations that struggle with security from those that thrive despite evolving threats.

Key Takeaways

- **Legal and regulatory compliance** requirements often drive security programs; duc care and due diligence are legal expectations for responsible security.
- **Information security governance** provides top-down strategic direction and requires active support from senior management to be effective.
- **Patch management** prevents the exploitation of known vulnerabilities through a structured process of testing, deployment, and verification.
- **Change management** ensures that all IT changes are reviewed and approved, preventing new security risks or operational instability.
- **A formal IT risk management program** helps organizations proactively identify, assess, and prioritize risks to make informed security decisions.
- **Personnel policies** establish clear expectations and accountability for employee behavior, covering everything from acceptable use to remote work security.
- **Security awareness and training** are essential for building a strong layer of defense by teaching people to recognize threats and follow security best practices.
- **A strong security culture**, where security is everyone's responsibility, is one of the most powerful security controls an organization can have.

Quick Action: Share Security Knowledge with Your Family

In this chapter, you learned about security management and culture, including how organizations build security awareness through training and create environments where security becomes everyone's responsibility. So in this quick action exercise, you'll apply these same principles by sharing what you've learned with someone in your family, whether that's a parent, spouse, sibling, or any other family member.

What to Do

Choose a Family Member to Help:

- Think about someone in your family who could benefit from better security practices. This could be someone you live with or a relative you stay in touch with remotely.
- Reach out and offer to spend 10–15 minutes helping them improve their security (frame it as something you're learning about and want to share, not as criticism of their current practices).

Share One Key Security Concept:

- Pick one topic from this book that you think they would benefit from the most, such as recognizing phishing emails, creating strong passwords, enabling multi-factor authentication, or securing their home Wi-Fi.
- Explain it in simple terms and why it matters for protecting their personal information and accounts.

Help Them Take One Concrete Action:

- Walk them through one practical security improvement, such as enabling MFA on their primary email account, showing them how to spot phishing emails, or verifying their devices have automatic updates enabled.
- Make sure they understand what you did together and why it helps protect them.

Note: If they're not interested right now, that's okay. You've planted a seed, and they may come back to you later. The goal is to make security feel helpful, not scary or judgmental.

By helping a family member improve their security, you've practiced the same security awareness and culture-building that organizations use to strengthen their overall security posture. You've also made someone you care about safer in the digital world.

Chapter 11: Application Security, Resilience, & Modern Devices

Introduction

Chapter 10 covered how organizations create the management structures, policies, and culture that support security. We looked at governance frameworks, risk management programs, personnel policies, and training, which are the organizational foundation that makes security possible. But frameworks and policies alone don't protect systems. Someone has to build secure applications, design resilient infrastructure, and secure the devices employees actually use. We now shift from management to implementation: the practical work of building and protecting technology in today's complex environment.

Modern security faces three major challenges that we'll explore in this chapter:

1. **Building Secure Applications**: Software vulnerabilities remain some of the most common attack vectors. If organizations are developing applications, whether for internal use or for customers, they need to build security in from the beginning.
2. **Withstanding Disruptions and Recovering Quickly**: This concept, called cyber-resiliency, recognizes that perfect prevention is impossible. Instead of asking "How do we stop all attacks?" organizations must ask "How do we keep operating when something goes wrong?" Whether it's a cyberattack, a power outage, or a natural disaster, organizations need the ability to withstand disruption, recover quickly, and adapt to new threats.
3. **Securing an Increasingly Diverse Technology Landscape**: Employees now commonly work from smartphones and tablets, organizations deploy a variety of Internet of Things (IoT) devices, and applications commonly run in virtual environments. Each of these technologies brings unique security challenges that traditional approaches don't fully address.

These three challenges, application security, cyber-resiliency, and securing modern devices, represent the practical reality of implementing security in today's organizations. Let's explore each one.

Application Development Security

Previous chapters focused on managing and securing existing IT systems. But what about the software that organizations develop themselves? When building custom applications, whether for internal use or for customers, how do you ensure those applications are secure? Application development security addresses this gap. It's about building security into software from the beginning, rather than trying to add it on later.

The Problem with Insecure Software

Historically, security wasn't a major focus in software development. Developers primarily focused on functionality, making software that worked and met business requirements. Security was often an afterthought, if it was considered at all.

This led to software with serious security flaws: Applications that transmitted passwords in plain text, were vulnerable to SQL injection attacks, had buffer overflow vulnerabilities, or failed to properly encrypt sensitive data. These weren't just theoretical problems; insecure software has been the root cause of countless security breaches.

I witnessed this problem firsthand as a web developer in the early 2000s. At that time, most websites used unencrypted HTTP for logins, sending passwords across the Internet in plain text. This wasn't just small sites; major platforms like Facebook didn't make encryption mandatory until the early 2010s. This practice turned public Wi-Fi at coffee shops and airports into hunting grounds, where attackers could use freely available tools to intercept passwords from unsuspecting users.

These insecure practices illustrate a critical point: insecure applications can bypass even the strongest network defenses. If an application sends passwords in plain text, no firewall can protect that data once it leaves the network.

> **KEY POINT**: While the widespread adoption of HTTPS has fixed this specific historical flaw, the core lesson remains: software vulnerabilities are a major problem, and modern security practices require building security in from the beginning, not adding it on as an afterthought.

How Software Is Built: A Quick Overview

Before we discuss how to build secure software, it helps to understand how software development typically works. Organizations use different development methodologies to structure their work.

Figure 11.1: Traditional Waterfall Model

- **Waterfall Model**: This traditional approach follows a linear sequence: requirements, design, development, testing, deployment, and maintenance, where each phase must be completed before the next begins. Think of it like building a house: you complete the foundation before framing the walls, and finish the walls before adding the roof.
- **Agile Model**: A more flexible approach that develops software in short cycles called "sprints," typically lasting 2–4 weeks. Each sprint delivers a working increment of the software, enabling teams to adapt to changing requirements and receive faster feedback.
- **DevOps Model**: Integrates development and IT operations teams to enable rapid, continuous deployment of software. Instead of large, infrequent releases, DevOps teams might deploy updates multiple times per day.

Each approach has strengths and weaknesses, but they all share a common security challenge: if security isn't built into the development process itself, you end up with insecure software.

The Secure Software Development Lifecycle (SDLC)

The solution to insecure software is to integrate security into the software development process itself. This is called the Secure Software Development Lifecycle (Secure SDLC or S-SDLC). This process adds security considerations to each phase of development:

- **Phase 1 - Requirements**: Security needs are identified upfront alongside business features. For example, a requirement might be, "The application must encrypt all customer data."
- **Phase 2 - Design**: Security is built into the application's architecture. This is where decisions are made about how to handle authentication, encryption, and protection against common attacks before any code is written.
- **Phase 3 - Development**: Developers write code following secure coding practices to prevent common vulnerabilities.

- **Phase 4 - Testing**: Security experts test the application for vulnerabilities alongside traditional quality assurance testing.
- **Phase 5 - Deployment**: The application is deployed with secure configurations, not insecure defaults.
- **Phase 6 - Maintenance**: Security remains a focus after release through ongoing monitoring and patching new vulnerabilities as they're discovered.

The key principle is that security is "baked in" throughout the development process, not "bolted on" at the end. It's much easier and more effective to build security in from the beginning than to try to add it to finished software.

Security Testing in Development

Testing is a critical part of the Secure SDLC. There are two main types of security testing that work together:

- **Static Code Analysis** tests code without running it. Think of it like editing a manuscript for grammatical errors before it's published. Security experts or automated tools review the source code, looking for patterns that indicate vulnerabilities. This allows developers to fix issues immediately, during the development phase.
- **Dynamic Testing**, by contrast, tests code while it's running. This is like giving the finished book to a group of readers and critics to see if they find any inconsistencies you missed. Testers execute the application and try to attack it, simulating what a real attacker would do. This reveals vulnerabilities that only appear when the code is actually running.

These two approaches complement each other. Static analysis can quickly identify certain types of vulnerabilities early in development. In contrast, dynamic testing finds vulnerabilities that only appear when the code is actually running and can validate that security controls work as intended.

TECH TALK: DevSecOps

DevSecOps is a newer approach that integrates security directly into the rapid development and deployment pipeline used in DevOps. Instead of a final security check, security becomes an automated, continuous part of the process. Tools automatically scan code for vulnerabilities every time a developer makes a change, providing immediate feedback. This approach is known as "shifting security left," which involves moving security from the end of the process to the earliest possible stages, like design and coding. This helps catch issues early and makes security a shared responsibility, not a final roadblock.

Authorization to Operate (ATO)

Before an application can be deployed into production (the live environment where real users access it), it often needs formal approval through a process called Authorization to Operate (ATO). This is especially common in government and regulated industries. An authorizing official (typically a senior executive) reviews a security package that documents the application's security controls, testing results, and identified risks. Based on this review, they decide whether the application's security is adequate. Think of an ATO as similar to how a building inspector must certify that a new building meets safety code standards before people can occupy it. The ATO serves as a formal acknowledgment that the application has been evaluated, its security risks are understood, and organizational leadership accepts those risks.

Why Application Security Matters

The true importance of application security lies in a crucial shift in mindset: moving from reactive to proactive security. For years, security was often an afterthought. Organizations would release software and simply react to security flaws as they were discovered—often after a breach had already occurred. This reactive approach is the story behind many of the major breaches we've studied. Remember the Sony breach (Chapter 6), caused by a classic SQL injection flaw that should have been prevented during development, or the Equifax breach (Chapter 1), which stemmed from a failure to reactively patch a known vulnerability in time? These disasters highlight a critical lesson: network defenses alone can't protect you if the applications themselves are insecure.

A proactive approach, using a Secure SDLC, flips this model on its head by building security in from the start to prevent flaws from reaching production. However, a mature program also recognizes that no proactive process is perfect; zero-day attacks can exploit unknown weaknesses. That's why strong reactive defenses, like the bug bounty programs we covered earlier, remain an essential safety net for finding and fixing vulnerabilities that inevitably slip through the cracks.

Ultimately, this is why application security matters: it's about creating a resilient process that's proactive by design but also prepared to react intelligently. By combining a secure development lifecycle with a robust plan for finding and fixing flaws after release, organizations build applications that aren't only developed securely but can also stay secure over time.

> ! **COMMON MISTAKE**: Treating security as something to add at the end of development. By the time an application reaches the testing phase, fixing security flaws requires significant rework. Building security in from the requirements and design phases is far more effective and less expensive.

Cyber Resiliency: Ensuring Business Continuity

So far, we've focused on protecting against attacks and preventing security breaches. But security isn't just about stopping bad guys; it's also about ensuring your organization can continue operating even when things go wrong. This is the domain of cyber resiliency.

Understanding Cyber Resiliency

Cyber resiliency is an organization's ability to withstand, recover from, and adapt to cyberattacks or other disruptions while continuing to deliver essential services. It represents a fundamental shift in thinking: from "How do organizations prevent all attacks?" to "How do they maintain operations even when attacks succeed?"

This shift in thinking matters because perfect prevention is simply impossible. Even with the best security controls, determined attackers may eventually succeed, systems can fail, and natural disasters happen. Since it's a matter of when, not if, a disruption occurs, cyber-resiliency determines whether your organization faces a minor hiccup or a catastrophic failure.

Cyber resiliency rests on three core principles:

- **Withstand**: Designing systems that can continue operating even during an attack or disruption. This might mean having redundant systems so that if one fails, others take over, or segmenting your network so that a breach in one area doesn't compromise everything.
- **Recover**: Establishing processes to quickly restore normal operations after a disruption. This includes having current backups, documented recovery procedures, and teams that know their roles during an incident.

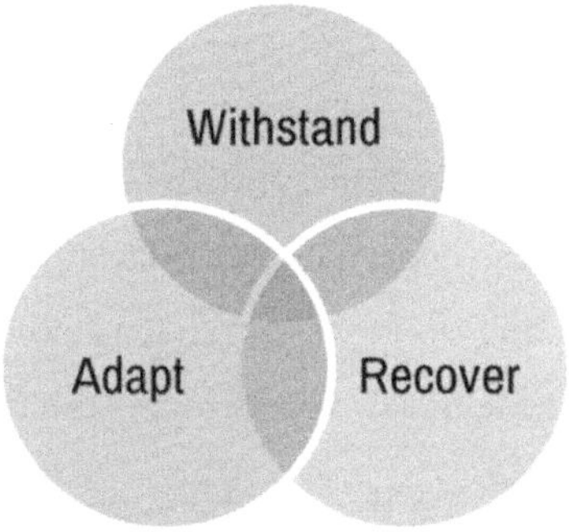

- **Adapt**: Learning from each incident to improve our defenses and response capabilities. After every disruption, organizations ask, "What can we do better next time? How do we prevent this from happening again?"

These principles tie directly to the CIA triad we discussed earlier in the book. While confidentiality and integrity focus on protecting data from unauthorized access and modification, availability (the "A" in CIA) requires resiliency to ensure systems remain accessible even during disruptions.

Imagine this scenario: Your company's data center is located in a single building. One night, a fire breaks out. The fire suppression systems activate and prevent major damage, but the data center needs to be shut down for inspection and repairs. If all your IT systems are in that one location, your business operations come to a halt. Employees can't access the systems they need to do their work. Customers can't access your services. And the business loses money every hour the systems are down. This is why organizations implement strategies like geographic dispersal and redundancy to ensure that localized disruptions don't bring down the entire business.

Geographic Dispersal: Don't Put All Your Eggs in One Basket

Geographic dispersal is the practice of spreading your IT resources across multiple physical locations. The idea is simple: if something happens to one location, your other locations can continue operating.

This applies to several types of resources:

- **Data Centers**: Instead of having all your servers in one data center, you might have data centers in different cities or regions. If one data center experiences problems, others can take over its workload.
- **Backups**: Your backup data should be stored in a different location from your primary systems. If you only keep backups in the same building as your primary systems, a disaster that affects the building could destroy both your primary data and your backups.
- **Cloud Services**: Cloud providers typically operate data centers in multiple geographic regions. When you use cloud services, you can configure your systems to use multiple regions. This means if one region experiences an outage, your systems can automatically failover to (switch to) another region.

The level of geographic dispersal you need depends on your organization's size, risk tolerance, and regulatory requirements.

Redundancy: Having Backups for Your Backups

While geographic dispersal tells you where to put your resources, redundancy is about what you put there: duplicate systems designed to take over if a primary system fails. Think of it as having a spare tire in your car. If you get a flat, you can replace it with the spare and continue your journey. Let's explore the different types of redundancy that work together to create a resilient infrastructure:

Data Redundancy: Protecting Your Most Valuable Asset

Data is often your organization's most valuable asset, and losing it can be catastrophic. Without access to customer records, financial information, and operational data, a business can grind to a halt. This is why data redundancy is a foundational part of resiliency. It ensures you have multiple copies of your data so that a single event, whether it's a hardware failure or a ransomware attack, never stops your business. Organizations use different backup strategies, each with distinct purposes:

- **Full Backups** copy everything. It's like taking a complete photograph of all your data at a specific moment in time. Full backups take the longest time and use the most storage space, but they're the most comprehensive. Restoration is straightforward—you simply restore the full backup.
- **Incremental Backups** only copy data that's changed since the last backup (whether that was a full backup or another incremental backup). They're fast and use minimal storage. However, restoration takes longer because you need the last full backup plus every incremental backup since then.
- **Differential Backups** copy everything that's changed since the last full backup. They're a middle ground—faster than full backups but easier to restore from than incremental backups. To restore, you only need the last full backup plus the most recent differential backup.

Most organizations combine these approaches, performing a full backup weekly, with either incremental or differential backups daily. This balances storage costs, backup time, and recovery speed.

But where should you store these backups? The **3-2-1 backup rule** provides a simple, memorable framework:

- **3 Copies of Data**: Maintain three copies: one primary and two backups.
- **2 Storage Locations**: Store copies in two separate locations (e.g., one on-premises, one offsite/cloud).
- **1 Off-Site Copy**: Ensure at least one backup is off-site to protect against local disasters.

This rule protects against multiple failure scenarios. If your building burns down, you have an off-site copy. If ransomware encrypts your primary systems and your local backups, you have a cloud copy.

System Redundancy: Keeping Services Running

While data backups protect your information, they don't keep your services online. If a web server crashes, customers can't access your website. If a database server fails, your applications stop working. This is where system redundancy comes in. System redundancy ensures that if a server or service fails, another immediately takes over, often so quickly that users never notice the disruption.

Load Balancing is a key technology for system redundancy. Instead of a single server handling all requests, you have multiple servers, along with a load balancer that distributes incoming requests across them. Think of it like a grocery store with multiple checkout lines. If one cashier is busy, customers can go to another line, and the store keeps operating smoothly.

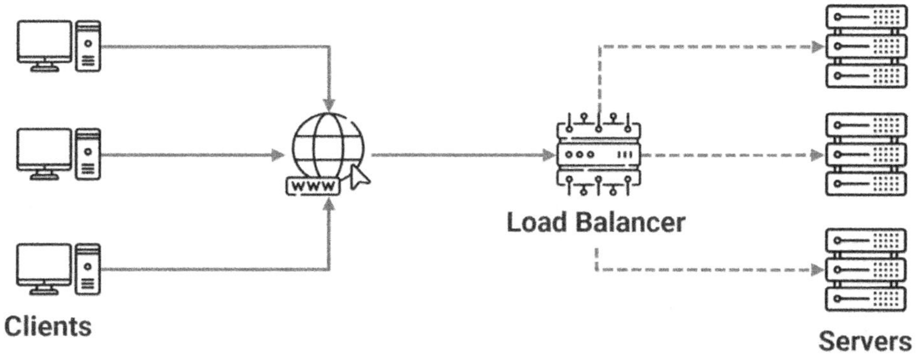

Figure 11.2: Load Balancing Diagram

Load balancing provides two critical benefits. First, it improves performance by distributing work across multiple servers. Second, it provides fault tolerance. If one server fails, the load balancer automatically stops sending requests to it and distributes the workload among the remaining servers.

For example, suppose a popular e-commerce website uses load balancing with five web servers. What would happen if one of the servers developed a hardware problem? The load balancer would detect that it's not responding and stop sending requests to it. The remaining four servers would continue handling customer traffic, where customers never experience an outage, and IT staff can fix the failed server without pressure.

Power Redundancy: When the Lights Go Out

Even the most resilient IT infrastructure stops working without electricity. Think about a hospital. During a power outage, critical life-support systems can't just shut down; they need a continuous power source. In the same way, an organization's critical IT systems need to remain online, as an abrupt shutdown can cause data corruption, failed transactions, and service outages. Power redundancy ensures systems continue to run when grid power fails.

- **Uninterruptible Power Supply (UPS)** units provide immediate battery backup when power fails. If a computer or server loses power abruptly, the operating system and any open files can become corrupted. A UPS acts as a short-term bridge, keeping critical systems running for a few minutes, which is just enough time to either shut them down gracefully or for the backup generators to start up.

- **Backup Generators** provide longer-term power during extended outages. These are typically large diesel or natural gas engines that can power an entire data center for days or even weeks, as long as they have fuel. They're the long-term solution that takes over after the UPS has bridged the initial gap.

You'll commonly see a combination of both in the business world, where UPS units provide an instant bridge for critical systems until a large backup generator kicks in to power the entire building. These principles scale down, too. In fact, for my own home office, I apply the same logic: I have three smaller UPS systems protecting my servers, networking equipment, and desktop, and a portable backup generator ready for longer outages to minimize downtime for my business.

Site Redundancy: When Location Matters

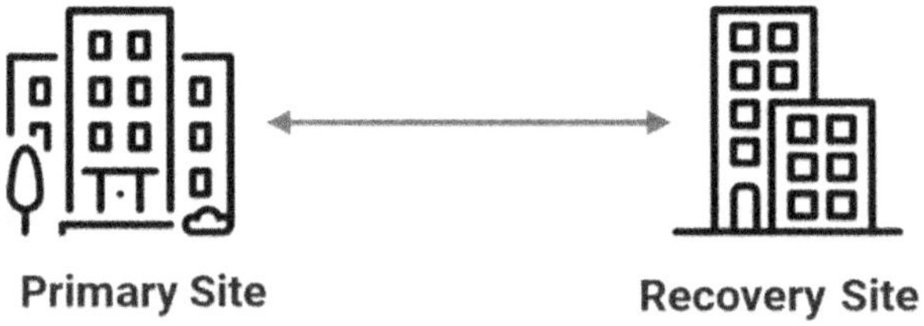

Figure 11.3: Site Redundancy

Sometimes, a disruption is bigger than a power outage; an entire facility might become unavailable due to a fire, flood, or other disaster. For these scenarios, you need to recover operations at an entirely different location. Site redundancy means having alternate facilities where operations can be restored:

- **Hot sites** are mirror images of your primary data center. They're fully equipped alternate facilities with all critical servers and networking, always running, with your data continuously replicated to them. Think of it as a fully furnished twin apartment, ready for you to move into at a moment's notice. If your primary site fails, you can switch to the hot site within minutes or hours. Because you're essentially running duplicate infrastructure, hot sites are the most expensive option, but they provide the fastest recovery.
- **Warm sites** are partially equipped facilities that can be brought online in hours to days. They have the basic infrastructure like power, cooling, and network connectivity, but you need to bring in your main servers and restore your data from backups. Think of it as an unfurnished apartment; the space

is ready, but you have to move in all your furniture. Warm sites offer a middle ground between cost and recovery speed.

- **Cold sites** are little more than an empty room. They're empty facilities with little to no pre-existing infrastructure where you can set up systems if needed, but you have to bring in and set up all of your equipment from scratch. Even basic utilities like power, cooling, and network connectivity may need to be installed. Recovery can take days to weeks. Cold sites are the least expensive option but provide the slowest recovery.

Testing Your Resilience

Having redundant systems and backup sites doesn't guarantee they'll work when you need them. The only way to know if your resilience plans will work is to test them regularly.

Tabletop Exercises bring together key personnel to walk through disaster scenarios on paper. A facilitator presents a scenario, such as "A ransomware attack has encrypted all data on our primary file servers," and the team discusses how they would respond. What's the first step? Who makes decisions? What resources are needed? Where are the gaps in our plan?

Tabletop exercises are relatively quick and inexpensive, and they often reveal problems that weren't obvious when the plan was written. Maybe the backup restoration procedure assumes a specific person will be available, but what if they're on vacation? Maybe the disaster recovery contact list has outdated phone numbers. These exercises surface such issues before a real disaster strikes.

Checklist Reviews verify that your documented procedures are up to date and complete. Someone physically walks through the actual steps in your disaster recovery checklist, verifying that every piece of information is accurate and every required resource is available. Are the backup tapes actually where the checklist says they are? Do the network configuration details match your current infrastructure?

Many organizations also conduct actual failover tests, switching operations to their backup site to verify everything works. These tests reveal technical problems that tabletop exercises might miss, but they're also riskier. Since you're testing live systems, a problem during the test could disrupt normal operations.

Testing reveals two types of problems: issues with your plans and issues with your implementation. Both matter. A perfect plan that nobody knows how to execute doesn't help. Neither does a plan that everyone knows but is based on outdated information.

The Security Connection

You might wonder why we're discussing concepts like backups and power generators in a cybersecurity book. The connection is simple: resiliency is a direct defense for the "Availability" pillar of the CIA triad. Many security incidents are designed specifically to disrupt availability. As we've learned, ransomware attacks make your data unavailable by encrypting it, and denial-of-service (DoS) attacks make your systems unavailable to legitimate users. In all these scenarios, your resiliency measures are what allow you to maintain operations and recover. If ransomware encrypts your primary systems, you restore from your offsite backups. If a DoS attack targets one data center, you can redirect traffic to another. Simply put, resiliency is the safety net that ensures your organization can survive and recover from a successful cyberattack.

Securing Modern Devices

The traditional model of workplace computing was simple: The company provided each employee with a company-issued desktop or laptop, most employees worked in a corporate office, and the IT department controlled and secured those computers. But the last decade has dramatically changed this landscape. Employees now often work from smartphones and tablets, organizations deploy a variety of IoT devices throughout their facilities, and many applications now run in virtual machines (VMs) and containers in the cloud, where each of these technologies brings unique security challenges.

To protect this modern landscape, we first look at the devices that are closest to us: the personal smartphones and tablets employees now use for work. This trend, known as BYOD, created the first major shift away from the traditional security model.

BYOD: Bring Your Own Device

The shift to the modern landscape began with BYOD, which stands for Bring Your Own Device. This refers to the practice of allowing employees to use their personal devices (smartphones, tablets, and sometimes laptops) for work purposes. Instead of carrying two phones (one personal and one for work), an employee uses their personal smartphone to access work email, company applications, and business data. BYOD has become popular for several reasons:

- Employees prefer using devices they're already familiar with and have chosen themselves.
- It can save companies money on hardware purchases.
- Employees can be more productive when they can work from their personal devices anywhere, anytime.

But BYOD creates significant security challenges. When employees use personal devices for work, how do you protect company data on devices you don't own and can't fully control?

The BYOD Security Challenge

Consider the security issues that arise when personal and professional worlds collide on the same device:

- **Loss of Control**: The device is the employee's personal property. Therefore, the company can't fully dictate what apps they install, what websites they visit, or how they use the device in their personal life. However, a BYOD agreement may prohibit the installation of certain high-risk applications.
- **Data Co-mingling**: The same device contains personal photos and messages right alongside confidential company emails and documents. How do you protect company data without invading the employee's privacy?
- **Physical Security**: Personal devices are easily lost or stolen. If a lost phone contains company data, it becomes a security incident.
- **Inconsistent Security**: An employee might not keep their device updated, may use a weak passcode, or might install apps from untrusted sources, creating security holes.
- **Offboarding**: When an employee leaves, how do you remove company data without wiping their personal photos and files?

The central challenge of BYOD is that organizations must protect sensitive company data on devices they don't own or fully control. These challenges mean that organizations need specific security controls for BYOD.

Mobile Device Management (MDM)

Mobile Device Management (MDM) is software that allows organizations to manage and secure mobile devices, including BYOD devices used for work. Think of it as a way for the IT department to apply security policies to devices they don't own.

With MDM, the employee installs a management app on their device. This app gives the IT department important security capabilities. They can require a passcode or biometric authentication to unlock the device and enforce encryption of all data stored on it. The system ensures devices stay up to date with security patches. If a device is lost, IT can remotely locate it, and if necessary, remotely wipe company data from it—without touching the employee's personal photos or files. The MDM can also prevent certain risky behaviors, like jailbreaking, which removes built-in security restrictions.

The key is that MDM provides security controls while respecting that it's a personal device. The IT department can enforce security requirements, but they can't access the employee's personal data or control their personal use of the device.

Making BYOD Work

Successful BYOD programs balance security with usability and privacy. If security controls are too restrictive or invasive, employees will resist using their personal devices for work, or they'll find ways around the controls. Conversely, if controls are too lax, company data is at risk.

A formal BYOD agreement is essential for setting clear expectations. This document outlines what the company is allowed to do via MDM and what the employee is prohibited from doing. The agreement should respect employee privacy by limiting controls to what's necessary for security, making it easy for employees to enroll their devices and comply with security requirements, providing support when employees have issues with their devices, and establishing clear policies about what happens when devices are lost or when employees leave.

BYOD isn't right for every organization. Some organizations, particularly those handling highly sensitive data or subject to strict regulations, may decide the risks are too great and prohibit BYOD entirely. Others embrace it as a way to support flexible work arrangements and employee preferences.

Mobile Device Hardening

Whether devices are company-owned or personal (BYOD), they need to be properly secured. Mobile device hardening refers to the security configurations and controls applied to smartphones and tablets to protect them against threats. Modern mobile devices face unique security challenges; they're portable and easily lost or stolen, they connect to untrusted networks, users install apps from various sources, and they contain both personal and business data. Effective mobile device hardening addresses these challenges.

Authentication and Access Controls

The first line of defense is strong authentication. Organizations should enable biometric authentication, such as fingerprint or face recognition, along with a strong passcode or PIN. Devices should be set to auto-lock after a short period of inactivity, typically 5 minutes or less for devices that access sensitive data. This ensures that if someone walks away from their device, it doesn't remain unlocked and accessible.

System and Application Management

Keeping the operating system current is critical. Devices should be configured to automatically install security updates. Mobile operating systems regularly receive patches for newly discovered vulnerabilities, and devices running outdated OS versions are vulnerable to attacks that patches have already fixed.

Application control is equally important. Organizations can restrict which apps can be installed or implement an approved app list. Consumer app stores contain both legitimate and malicious apps disguised as legitimate ones. Some organizations implement their own enterprise app stores with vetted applications to avoid this risk. Organizations also need systems that detect jailbroken iOS devices or rooted Android devices, since these processes remove built-in security restrictions. Jailbroken devices can bypass security controls and should be blocked from accessing corporate data.

Network Security

Finally, network security requires VPN connections when accessing corporate resources over untrusted networks. Organizations should prohibit or restrict the use of public Wi-Fi for sensitive work. Some implement a per-app VPN that automatically secures specific applications whenever they're used, regardless of the network connection.

EXAMPLE: A healthcare organization issues smartphones to doctors and nurses for accessing patient records. They enforce several hardening measures: biometric authentication with a 6-digit PIN backup, automatic installation of OS updates, blocking of jailbroken devices, and an automatic VPN connection when accessing the patient database. If a device is lost, it's configured to auto-wipe after 10 consecutive failed authentication attempts, protecting against brute-force attacks while allowing for legitimate mistakes. These layered controls ensure that a lost device doesn't become a HIPAA violation.

Internet of Things (IoT) Security

The IoT refers to physical devices embedded with sensors, software, and network connectivity—devices that weren't traditionally thought of as "computers" but now connect to networks and exchange data. Think security cameras, smart thermostats, building access systems, industrial sensors, smart TVs, and connected medical devices.

IoT devices have become ubiquitous in organizations, but they create unique security challenges that traditional IT security approaches don't fully address. The scale of this problem is staggering. IoT breaches soared to 1.51 billion in just the first six months of 2022 (a 77% increase from the previous year). That's roughly 8.4 million breaches every single day.

Why IoT Security Is Different

IoT devices present unique challenges that distinguish them from traditional IT equipment:

- **Limited Built-in Security**: Many IoT devices are designed for functionality and cost-effectiveness rather than security. They often lack basic security features such as strong authentication, encryption, and the ability to install security updates. Additionally, some come with hardcoded default passwords that can't be changed.
- **Long Lifespans**: Unlike computers, which are often replaced every few years, IoT devices such as security cameras and building sensors often remain in service for many years or even decades. A security camera installed in 2015 might still be in use in 2025, but it likely hasn't received security updates for years.
- **Always Connected**: IoT devices typically maintain constant network connections, providing a persistent attack surface. A compromised IoT device can serve as a foothold for attackers to access other systems on your network.
- **Diverse Ecosystems**: Organizations might have IoT devices from dozens of different manufacturers, each with its own management interface, security capabilities, and update mechanisms. This diversity makes consistent security management difficult.

EXAMPLE: Mirai Botnet (2016)

In 2016, the Mirai botnet compromised hundreds of thousands of IoT devices, primarily security cameras and home routers, by using default passwords that users had never changed. The botnet then launched massive distributed denial-of-service attacks that disrupted major Internet services. The devices weren't targeted because of what they did; they were targeted because they were poorly secured and always connected.

Essential IoT Security Controls

Given these challenges, organizations need specific strategies for securing IoT devices. The following controls form the foundation of a strong IoT security program:

Network Segmentation: Isolating the Risk

IoT devices should be placed on isolated network segments, separate from your main business network. Think of it like having separate keys for different areas of a building—the maintenance staff key opens the utility rooms but not the executive offices. Similarly, IoT devices should only access the specific network resources they need, nothing more. A compromised security camera shouldn't be able to access your file servers or financial systems.

Secure Configuration from Day One

When deploying IoT devices, security must begin immediately. All default usernames and passwords should be changed before the device ever connects to your network. Many IoT compromises happen because devices are left with factory default credentials like admin/admin or admin/password. Each device needs a strong, unique password.

Beyond passwords, any features or services the device doesn't need should be turned off. If you're using a smart TV for digital signage in your lobby and it doesn't need to access the Internet, disable its Internet connection. If a device has unnecessary network ports or protocols enabled, disable them. Every active feature is a potential attack vector.

Keep Firmware Updated

Organizations should establish processes for regularly updating IoT device firmware. Unlike computers that often update automatically, many IoT devices require manual firmware updates that administrators must initiate. This means creating an inventory of all IoT devices, tracking their firmware versions, and scheduling regular updates. Yes, this is time-consuming, but it's essential. Unpatched IoT devices are low-hanging fruit for attackers.

Monitor for Suspicious Behavior

Network monitoring should be deployed to detect unusual behavior from IoT devices. Your security camera should be sending video to your recording system; that's normal. But if that same camera suddenly starts scanning your network for other devices or sending gigabytes of data to an unknown server in another country, that's a clear sign of compromise. Monitoring tools can detect these anomalies and alert your security team before a small breach becomes a major incident.

> ! **COMMON MISTAKE**: Treating IoT devices as "set it and forget it" technology. Many organizations deploy IoT devices and never think about them again until something goes wrong. Successful IoT security requires ongoing management: tracking devices, monitoring behavior, applying updates, and decommissioning or replacing devices that no longer receive security support from the manufacturer.

Securing Virtualized Environments

Modern IT infrastructure increasingly relies on virtualization, a technology that allows multiple virtual systems to run on a single server. Understanding how to secure these virtualized environments is essential for protecting modern organizations.

Virtual Machines (VMs)

VMs are software-based emulations of physical computers. Think of a VM like an apartment in a building. The building (physical server) provides the foundation, utilities, and structure. The software that creates and manages these apartments is called a hypervisor; you can think of it as the building manager who ensures each apartment gets the resources it needs and stays separate from the others. Each apartment (VM) is a complete, self-contained living space with its own operating system and applications. One apartment might have completely different furniture, décor, and residents than another, but they all share the same building infrastructure.

A single physical server might run 10, 20, or even 50 VMs, each running its own operating system and applications. This efficiency is why virtualization has become the foundation of modern data centers and cloud computing.

Containers

Containers are a lighter-weight form of virtualization. Instead of including a full operating system like VMs do, containers package just the application and its specific dependencies.

Because they're so lightweight, containers start faster, use fewer resources, and are more portable than VMs. A VM might take minutes to start up, while a container can start in seconds. A VM might require gigabytes of disk space, while a container might need only megabytes. This makes them ideal for modern application architectures where you might need to quickly scale up or down, or move applications between different environments.

To understand this difference, we can build on our previous analogy. If a VM is like an apartment, then a container is like a hotel room. While each apartment has its own full set of utilities (a complete OS), all the hotel rooms share the building's central infrastructure (the host OS). This shared approach is what makes each hotel room (container) much lighter and faster to make available for use. Just as a hotel offers different room sizes, containers can be allocated different amounts of resources like CPU and RAM based on the application's needs.

Security Challenges in Virtual Environments

Virtual environments offer tremendous efficiency and flexibility, but they also introduce specific security considerations that don't exist in traditional physical infrastructure:

Shared Resources and Isolation Risks

Multiple virtual systems share the same underlying physical hardware: processors, memory, storage, and network interfaces. If isolation between these systems isn't perfect, one compromised VM or container might be able to access or affect others on the same physical host. Imagine a malicious actor who compromises a low-security development VM and uses it as a stepping stone to attack production VMs containing customer data—all running on the same physical server. This "neighbor attack" is a unique risk in virtual environments.

The Critical Foundation Layer

The software that manages VMs (called a hypervisor) or containers (called a container runtime) becomes a critical security component. If an attacker compromises the hypervisor or container runtime, they potentially compromise all the VMs or containers it manages. This creates what's called a "single point of failure." It's like a master key that opens every apartment in the building—if someone gets that key, they can access everything.

Configuration Complexity

Virtual environments offer extensive configuration options, and misconfigurations can create security vulnerabilities. For example, VMs might be configured with unnecessary network access, or containers might run with excessive privileges. The flexibility that makes virtualization powerful also makes it easy to accidentally create security gaps.

Virtual Sprawl

VMs and containers are easy to create, which can lead to "sprawl" where organizations lose track of how many VMs or containers they have running and what's in them. Forgotten or unmanaged virtual systems become security risks, especially if they're not being patched or monitored. It's like discovering you've been paying rent on storage units you forgot existed—except these forgotten systems might contain vulnerabilities that attackers can exploit.

Key Security Controls for Virtual Environments

Protecting virtualized environments requires a multi-layered approach that addresses both the unique risks and the scale of virtual infrastructure.

Secure the Foundation

The first priority is keeping the software that runs your virtual environments (hypervisors and container runtimes) up to date with the latest security patches. These components are the foundation of your virtual infrastructure, and a vulnerability here affects everything running on top of them. Think of it like maintaining the structural integrity of a building; if the foundation is compromised, every apartment inside is at risk.

It's also critical to enforce strong isolation between VMs and containers. Modern hypervisors and container runtimes include isolation features, but they must be properly configured to ensure effective isolation. This prevents a breach in one VM or container from affecting its neighbors on the same physical host.

Build with Security in Mind

Whether creating VMs or containers, start with minimal base images that include only the necessary components. Every additional piece of software is another potential vulnerability. Container images should be as minimal as possible. If your application doesn't need a component, don't include it. This principle of "least functionality" significantly reduces your attack surface.

Continuous Vulnerability Management

Implement regular scanning of VM and container images to identify vulnerabilities in the operating system, application software, and dependencies. Many container security tools can scan images before deployment and block those with critical vulnerabilities. This catches security issues before they ever reach your production environment.

Beyond scanning static images, deploy security monitoring that can detect malicious behavior in running VMs or containers. This runtime protection might include detecting unusual network connections, unexpected process executions, or attempts to access sensitive files. For example, if a container that normally communicates only with your database suddenly starts connecting to external command-and-control servers, that's a red flag that needs immediate investigation.

Manage the Entire Lifecycle

Implement proper lifecycle management for VMs and containers. Keep an inventory of what you're running: every VM and container, what they do, who owns them, and when they were last updated. Decommission systems when they're no longer needed. It's also a good practice to regularly rebuild VMs and containers from trusted images rather than patching old ones indefinitely. This "immutable infrastructure" approach ensures you're working with known-good configurations and eliminates the drift that happens when systems are patched over months or years.

The shift to virtualized environments doesn't eliminate security challenges; it transforms them. Traditional security tools that assume physical separation between systems don't work in environments where dozens of systems share the same hardware. Understanding and implementing security controls designed for virtual environments is essential for protecting modern infrastructure.

Bringing It All Together

This chapter explored the practical work of implementing security in today's complex world, shifting our focus from management frameworks to the technologies we build and protect. We covered three critical pillars: building secure applications, ensuring cyber-resiliency, and securing the diverse landscape of modern devices.

First, we learned that application security requires a proactive mindset. By integrating security into every phase of the Software Development Lifecycle (SDLC) and conducting thorough testing, we can build applications that are secure by design. This proactive approach prevents flaws like those that led to the Sony (Chapter 6) and Equifax (Chapter 1) breaches. It's far more effective than reactively patching vulnerabilities after they've been discovered.

Next, we explored cyber resiliency, which accepts that perfect prevention is impossible. It's the safety net that ensures your organization can withstand and recover from disruptions, whether they're cyberattacks like ransomware, hardware failures, or natural disasters. Through strategies like geographic dispersal, data backups, and system redundancy, resiliency directly supports the "Availability" pillar of the CIA triad, ensuring your business can continue operating when things go wrong.

Finally, we addressed the challenge of securing modern devices. The traditional security model no longer applies in a world of BYOD, IoT, and virtualized environments. Each of these technologies requires a tailored approach, from using MDM to protect data on personal phones, to isolating IoT devices on segmented networks, to securing the foundational hypervisors that run our VMs.

Together, these three elements show that modern security is about thoughtful design, layered defenses, resilient architecture, and continuous adaptation. It's about building security in from the start, being prepared to recover when incidents occur, and securing the ever-evolving technologies that power our organizations.

Key Takeaways

- **Application security requires a proactive mindset**. the SDLC is about "baking in" security from the start, which is far more effective than "bolting it on" at the end.
- **Cyber-resiliency is about ensuring availability when prevention fails.** It's the ability to withstand, recover from, and adapt to disruptions, recognizing that it's a matter of when, not if, an incident will occur.
- **The 3-2-1 backup rule is your data's best defense**. Keep three copies of your data on two different types of media, with one copy stored offsite. This protects you from ransomware, hardware failure, and physical disasters.
- **Untested resiliency plans are unreliable.** Regularly testing your backup and recovery procedures through tabletop exercises and checklist reviews is the only way to know they'll work during a real crisis.
- **BYOD requires managing data on devices you don't own,** and Mobile Device Management (MDM) is a key tool that allows organizations to enforce security policies on personal devices while respecting employee privacy.
- **IoT devices create unique risks and must be isolated**. Due to their limited security features, they should be placed on separate network segments to prevent a compromised camera or sensor from accessing critical systems.
- **Virtualization security starts at the foundation.** In virtual environments, the software that runs your VMs and containers (the hypervisor or container runtime) is a critical single point of failure. If it's compromised, everything running on it is at risk.

Quick Action: Protect Your Data with Backups

In this chapter, you learned about cyber resiliency—the ability to withstand and recover from disruptions. The foundation of resiliency is having reliable backups of your important data. So in this quick action exercise, you'll assess your current backup situation and take steps to protect your most valuable digital assets.

What to Do

Assess Your Current Backup Situation:

- Do you have at least one backup of your most important data (photos, financial records, etc.)?
- If you don't, you're at risk of losing everything to a hardware failure, ransomware attack, or theft.

Identify What Needs to Be Backed Up:

- Make a quick list of your most irreplaceable data: family photos, important documents, financial records, and anything else you can't afford to lose (tax returns, work documents, etc.).

Implement at Least One Backup Method:

- **Option A: Local backup**: Purchase an external hard drive and set up automatic backups using your computer's built-in backup software (Windows Backup or macOS Time Machine).
- **Option B: Cloud backup**: Sign up for a cloud storage service like Dropbox, Google Drive, OneDrive, or iCloud and configure it to automatically sync your important folders.
- **At a minimum, implement one of these methods.** Ideally, do both—this follows the resiliency principle we discussed: local backups are fast for recovery, while cloud backups protect you if your home is affected by fire, flood, or theft.

Test Your Backup:

- Once you've set up a backup, test it by restoring a single file to verify it works. Set a calendar reminder to check your backups once a month to ensure they're still running and your data is being protected.

By implementing a backup strategy, you've taken one of the most important steps in cyber resiliency. You're now protected against hardware failures, ransomware attacks, natural disasters, and accidental deletions.

PART 4 CONCLUSION

Bringing It All Together

You've now learned how to implement the defenses that stop the threats you studied in Part 3. You understand the essential security controls that protect identity, devices, and network communications, from multi-factor authentication and encryption to firewalls and VPNs. You've explored how security operations teams monitor networks continuously, detect threats in real-time, respond to incidents systematically, and proactively test defenses through security assessments. You've learned how organizations create the management frameworks, policies, and security-conscious culture that make these technical controls effective. And you've discovered how to build security into applications from the beginning, ensure business continuity when attacks succeed, and protect modern technologies like mobile devices, IoT devices, and virtualized infrastructure.

The case studies throughout these chapters, from TJX's wireless security negligence (Chapter 8) to the Target breach's security operations failures (Chapter 9), demonstrate that effective security requires more than just deploying tools. These breaches happened not because the organizations lacked security tools, but because they failed to implement and use them properly. Target detected the malware but didn't respond to alerts. TJX used outdated encryption that they knew was broken. The true lesson is that effective security requires operational vigilance, strong management support, proper configuration, regular testing, and a culture where everyone understands their role in keeping the organization safe. You've now completed your comprehensive introduction to cybersecurity fundamentals.

You understand the threats, the defenses, and how they interact. The final section addresses the question I know many of you have been asking throughout this journey: "How do I turn this knowledge into a career?" It's time to explore your path forward.

PART 5: YOUR PATH FORWARD

Overview

You've completed a comprehensive journey through cybersecurity fundamentals, from understanding threats and defensive principles to implementing specific security controls and building resilient systems. Now comes the practical question that may have been on your mind since you opened this book: "How do I actually get started in this field?" Chapter 12 provides your roadmap for breaking into cybersecurity, whether you're starting fresh in IT or transitioning from another field. You'll learn about the diverse career opportunities available, realistic expectations for breaking into the field, common entry-level roles and their requirements, foundational certifications that help you stand out, and practical ways to gain hands-on experience even before landing your first security job. Chapter 13 wraps up your journey with final guidance and encouragement as you continue building your expertise. The cybersecurity field desperately needs talented, dedicated professionals. The global shortage of millions of qualified security practitioners means there has never been a better time to enter this field. Your journey doesn't end with this book—it begins right now.

Chapter 12. Getting Started in Cybersecurity

Introduction

If you've made it this far in the book, you now have a solid foundational understanding of cybersecurity: what it is, why it matters, how hackers operate, and the essential technical concepts and security controls that protect organizations from cyberthreats. You understand the CIA triad, defense-in-depth, network security, cryptography, access control, and the various layers of protection that work together to create a robust security posture. You've journeyed from understanding threat landscapes and hacker methodologies to exploring malware, social engineering, security operations, risk management, and application security. Now it's time to take that knowledge and transform it into a career.

The natural next question is: How do I actually get started in cybersecurity? This chapter provides a practical roadmap for breaking into the field. We'll explore the diverse career opportunities available, set realistic expectations about what it takes to succeed, discuss common entry-level roles, review foundational certifications that can help you stand out, and highlight practical ways to gain hands-on experience, even if you don't have a job in cybersecurity yet.

The good news is that cybersecurity is one of the most in-demand fields in IT today. As we discussed earlier in the book, there's a global shortage of over 3 million qualified cybersecurity professionals, and organizations are struggling to fill critical roles. This creates a tremendous opportunity for newcomers willing to put in the work to learn and develop their skills. Let's explore how you can position yourself to take advantage of these opportunities.

The Diverse World of Cybersecurity Careers

One of the most important things to understand about cybersecurity is that it's an incredibly diverse field with many career paths. When people first think about cybersecurity, they often picture a hoodie-wearing ethical hacker performing penetration tests or vulnerability assessments—hacking into systems with permission to test an organization's security posture. While that's certainly one path, it represents only a small slice of the opportunities available.

As we covered in Chapter 1, cybersecurity offers diverse career paths. Let's revisit those roles and expand the list to include additional specialized positions you might encounter:

- **Security Analysts** who monitor networks and systems for threats
- **Incident Responders** who jump into action when breaches occur
- **Security Engineers** who design and implement technical security solutions
- **Security Architects** who develop comprehensive security strategies
- **GRC Analysts** who ensure compliance with regulations and manage risk
- **Cloud Security Engineers** who specialize in protecting cloud environments
- **Threat Intelligence Analysts** who research emerging threats
- **Malware Analysts** who reverse-engineer malicious software
- **Threat Hunters** who proactively search for hidden threats in the network

And many more specializations beyond these.

The point is this: Don't limit yourself to just one area when you're starting out. Explore different aspects of cybersecurity to discover what genuinely interests you. You might be surprised to find that you're more drawn to the defensive side (blue team) than offensive security (red team), or that you prefer working with policies and compliance rather than hands-on technical work. All of these roles are critical to an organization's security posture, and all offer rewarding career paths.

> **PERSONAL NOTE:** I never intended to work in cybersecurity. When I started graduate school, my goal was to earn an MBA in IT Management and move into leadership roles. But everything changed when I was accepted into the CyberCorps Scholarship for Service (SFS) program, a government initiative that funds cybersecurity education in exchange for public service. That experience shifted my focus to Information Assurance and Security Management. The scholarship opened doors I never expected, leading me to public service with a mission to protect our government and its citizens. I took an oath to serve my country and spent 12 years specializing in Governance, Risk, and Compliance at the U.S. Department of the Treasury. It wasn't a path I'd even considered when I started, but it became deeply meaningful and rewarding.

Here's what I learned from that experience: Your career path rarely unfolds exactly as planned, and that's okay. My varied IT background—web development, desktop support, systems administration, database work—became the foundation for everything I accomplished in cybersecurity. Understanding how systems, networks, and applications actually work gave me the context I needed to secure them effectively. Without that strong technical foundation, I'd have struggled in security roles. Keep an open mind as you explore different areas of cybersecurity; your perfect role might be something you haven't even discovered yet.

Setting Realistic Expectations

Before diving into the specifics, let's set some realistic expectations about what it takes to succeed in this field.

Cybersecurity isn't an entry-level IT field. While there are entry-level cybersecurity positions, most organizations prefer candidates who have some foundational IT knowledge and experience first. This makes sense when you think about it. How can you secure networks if you don't understand how networks function? How can you protect systems if you've never administered them? And how can you identify suspicious activity if you don't know what normal looks like?

For many people, the path into cybersecurity starts with a more general IT role, such as:

- Help desk or desktop support
- Systems administration
- Network administration
- IT support specialist

These roles provide the foundational knowledge of how IT infrastructure works, which becomes invaluable when you transition into security. Many professionals spend several years in general IT roles before moving into security positions. You'll learn how users interact with systems, how networks are configured, how operating systems function, and how to troubleshoot problems, all skills that directly translate to security work.

That said, it's possible to break directly into cybersecurity if you're willing to invest significant time in self-study, hands-on labs, and certifications. The key is demonstrating that you have the foundational knowledge and practical skills that employers are looking for, even if you haven't gained them through traditional IT work experience. While most organizations prefer candidates with IT experience, some do offer entry-level security positions, such as Security Analyst I or Junior SOC Analyst, to candidates with strong foundational knowledge, relevant certifications, and demonstrated hands-on experience.

Cybersecurity requires continuous learning. The threat landscape is constantly evolving. New vulnerabilities are discovered daily. Attack techniques change. New technologies emerge. What you learn today will need to be updated and expanded within months, not years. Successful cybersecurity professionals are committed to lifelong learning through formal training, certifications, self-study, and hands-on experimentation.

> KEY POINT: Even after decades in IT and cybersecurity, I maintain a diverse home lab for exactly this reason. Lifelong learning isn't just advice I give my students; it's how I continue to stay current and explore new technologies. My home lab—which includes multiple servers, VMs, networking equipment, and various security tools—allows me to experiment with new security tools, test different configurations, and keep my skills sharp. The learning never ends, and that's actually one of the most exciting aspects of this field.

It takes time to develop expertise. Don't expect to become a penetration tester or security architect overnight. These are advanced roles that require years of experience and deep technical knowledge. Be patient with yourself, focus on building a strong foundation, and recognize that your career will develop over time. Most successful cybersecurity professionals look back and realize their early "detours" through help desk or systems administration were actually essential preparation.

With these expectations in mind, let's look at the practical steps you can take to get started.

Common Entry-Level Roles

When you're first breaking into cybersecurity, you'll typically be looking at entry-level positions such as:

Security Analyst / SOC Analyst

This is one of the most common entry points into cybersecurity. Security analysts work in Security Operations Centers (SOCs), monitoring networks and systems for security threats, analyzing alerts from security tools like SIEM and EDR, investigating potential incidents, and escalating serious threats to more senior team members. This role gives you broad exposure to many aspects of cybersecurity and helps you understand how attacks unfold in real-world environments. As a Tier 1 analyst, you'll typically work as part of a 24/7 team, which may include shift work. You'll learn to distinguish between false positives and real threats, understand attack patterns, and develop the analytical skills that form the foundation of many other security roles.

Junior Incident Responder

Similar to a SOC analyst, junior incident responders focus specifically on responding to security incidents when they occur. You'll learn how to contain threats, investigate breaches, collect evidence, and support recovery efforts. This role is excellent for developing practical skills in a high-pressure environment. The experience you gain is invaluable; you'll see firsthand how attacks progress, how defenders respond, and what works (and doesn't work) in real incident scenarios.

Junior Penetration Tester / Ethical Hacker

For anyone interested in offensive security, junior penetration tester roles are available. However, they're highly competitive and typically require demonstrated hands-on skills with penetration testing tools and techniques. In this role, you'll perform authorized security testing against an organization's systems, networks, and applications to identify vulnerabilities before malicious hackers can exploit them. You'll need to show proficiency with security tools, an understanding of common vulnerabilities, and the ability to think like an attacker. This role requires continuous learning as you'll need to stay current with the latest attack techniques and security tools.

Security Operations Specialist

This role often involves supporting the security team with tasks like vulnerability scanning, patch management, security tool administration, and documentation. While it may not be as exciting as performing penetration tests, it provides valuable exposure to the operational side of cybersecurity and helps you understand how security programs operate day to day. You'll learn about the security tools organizations use, how they're configured and maintained, and how different security controls work together. This foundational knowledge is essential for advancing to more specialized roles.

GRC Analyst (Junior)

If you're more interested in the policy, compliance, and risk management side of cybersecurity, a junior GRC role might be a good fit. You'll work on security documentation, assist with compliance audits, help conduct risk assessments, and support the development of security policies and procedures. This role is ideal for people who enjoy working with frameworks and standards, have a strong attention to detail, and prefer a more structured approach to security. While it's less technical than other entry-level roles, it provides critical insight into how organizations manage security from a governance perspective and how security aligns with business objectives. GRC roles can also lead to management positions, as they develop the strategic and business thinking that senior leaders need.

IT Security Intern

Don't overlook internships, even if you're not currently a student. Many organizations offer internship programs that can provide valuable hands-on experience and potentially lead to full-time positions. Internships are an excellent way to get your foot in the door and learn from experienced professionals. You'll gain exposure to real-world security operations, build your professional network, and demonstrate your capabilities to potential employers.

The key to landing these entry-level roles is demonstrating that you have the foundational knowledge and skills required, which brings us to certifications and training.

Foundational Certifications

Certifications play an important role in cybersecurity careers. They validate your knowledge, demonstrate your commitment to the field, and help you stand out to employers, especially when you don't yet have extensive work experience. However, it's important to understand that certifications alone won't get you a job. They need to be combined with practical skills, hands-on experience, and the ability to apply what you've learned. Here are the most valuable foundational certifications for beginners:

CompTIA Security+

This is widely considered the leading entry-level cybersecurity certification. Security+ covers a broad range of security topics, including threats and vulnerabilities, security architecture, cryptography, identity and access management, risk management, and incident response. It's vendor-neutral, widely recognized, and required for many U.S. government positions. If you're going to pursue one certification to start your cybersecurity career, Security+ is often recommended as the strongest choice for most people. The exam tests both theoretical knowledge and practical application, ensuring you understand not just what security controls exist, but how and when to use them. Security+ is often the minimum requirement listed in job postings for entry-level security positions, making it an essential credential for breaking into the field.

ISC2 Certified in Cybersecurity (CC)

This is a newer certification designed specifically for absolute beginners. You don't need any prior IT or security experience to pursue it. The CC covers fundamental security concepts and provides a solid introduction to the field, making it an accessible starting point. If you're brand new to both IT and cybersecurity, the CC can be a good starting point before moving on to Security+. The certification covers security principles, business continuity, disaster recovery, access controls, network security, and security operations at a high level, giving you a strong foundational understanding needed to pursue more advanced certifications.

Blue Team Level 1 (BTL1)

This certification from Security Blue Team focuses specifically on defensive security operations, which is the blue team side of cybersecurity. It's hands-on and practical, covering topics like security fundamentals, phishing analysis, threat intelligence, digital forensics, SIEM, and incident response. BTL1 has gained recognition in the industry for providing practical skills that directly translate to SOC analyst and incident response roles. Unlike purely theoretical certifications, BTL1 requires you to actually perform security analysis tasks, which many employers value as it demonstrates practical capabilities. If you're interested in defensive security and want a certification that teaches real-world skills, BTL1 can be a strong option.

INE eJPT (Junior Penetration Tester)

The eJPT from INE is a well-regarded hands-on certification for aspiring penetration testers. The eJPT exam requires you to actually perform ethical hacking tasks in a lab environment, pivoting between networks, exploiting vulnerabilities, and completing specific objectives. It's more affordable than many other certifications and provides the practical skills needed to work as a junior penetration tester. The exam gives you 72 hours to complete a series of challenges in a realistic network environment, testing your ability to enumerate systems, identify vulnerabilities, exploit weaknesses, and achieve specific objectives. If you're interested in offensive security and want a certification that demonstrates you can actually perform penetration testing tasks, the eJPT is worth considering. It bridges the gap between theoretical knowledge and practical application, giving you the confidence to handle real-world penetration testing scenarios.

CompTIA CySA+ (Cybersecurity Analyst)

CySA+ focuses on threat detection, analysis, and response, and is designed for those pursuing SOC analyst, security analyst, or incident response roles. The certification covers behavioral analytics, threat intelligence, security monitoring, incident response procedures, and vulnerability management. It emphasizes the analytical skills needed to identify and respond to threats in real-time. You'll learn how to interpret data from security tools, identify indicators of compromise, understand attacker behavior, and respond effectively to security incidents. The exam includes performance-based questions that require you to demonstrate practical skills, not just recall facts. CySA+ can be pursued alongside or after Security+ and serves as an alternative to BTL1 for those interested in defensive security operations.

Other Foundational IT Certifications

If you're brand new to IT, these foundational certifications are highly recommended before or alongside your security certifications:

CompTIA A+

The A+ certification covers computer hardware, operating systems, mobile devices, networking basics, troubleshooting methodology, and security fundamentals. While it's not a security-specific certification, A+ provides the foundational knowledge you need to understand how computers and devices actually work. In cybersecurity, you'll frequently encounter situations where you need to understand hardware components, how operating systems function, and how to troubleshoot problems. Without this foundational knowledge, more advanced security concepts can be more difficult to grasp. A+ fills these knowledge gaps and provides the baseline IT literacy that employers expect. If you're completely new to IT and want to break into cybersecurity, starting with the A+ gives you the foundation you need to succeed. It's also an excellent credential for landing your first IT job in help desk or desktop support, which can serve as a stepping stone to security roles.

CompTIA Network+

Network+ is considered by many to be among the most important foundational certifications for anyone pursuing a career in cybersecurity. The Network+ covers networking concepts, infrastructure, network operations, network security, and network troubleshooting. You'll learn about the OSI model, TCP/IP, routing and switching, wireless networking, network devices, protocols, and much more. This knowledge is essential for cybersecurity work. When you're analyzing network traffic in a SIEM, configuring a firewall, or investigating a security incident, you need to understand how networks function. Network+ provides all of this foundational knowledge. Many cybersecurity professionals consider Network+ essential for building a strong foundation, as security is built on networking concepts.

Cloud Certifications

Cloud computing is widely adopted by organizations. Most organizations use cloud platforms like Microsoft Azure, Amazon AWS, or Google Cloud. Getting certified on one of these platforms can be valuable for those interested in cloud security roles. Consider foundational certifications like:

- **Microsoft Azure Fundamentals (AZ-900)** and **Azure Security Technologies (AZ-500)**
- **AWS Certified Cloud Practitioner** and **AWS Certified Security Specialty**

- **Google Cloud Digital Leader** and **Google Professional Cloud Security Engineer**

Cloud certifications demonstrate that you understand how cloud services work, how they're secured, and how to implement security controls in cloud environments. As more organizations migrate to the cloud, cloud security skills are increasingly in demand.

Choosing Your First Certification

Based on my experience teaching over 200,000 students from diverse backgrounds, from undergraduates interested in IT to mid-career professionals seeking a change, here are some key considerations for choosing certifications.

Your certification path depends on your current experience level and career interests. Use the visual guides below to find your path. Think of them as common routes, not rigid requirements. Find the point that aligns with your background, then progress toward your area of interest.

Finding Your Starting Point

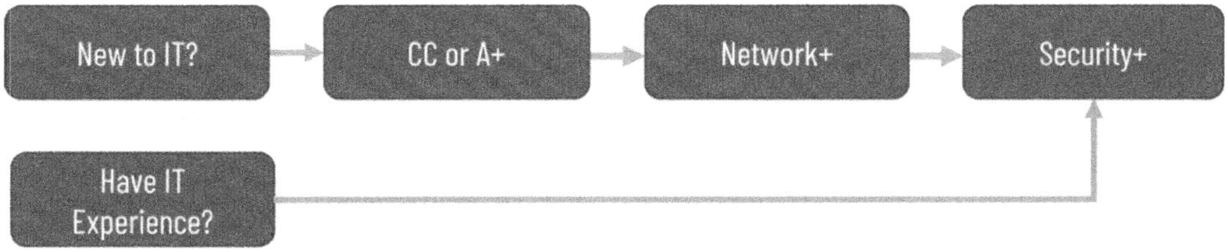

Figure 12.1: Finding Your Starting Point

Understanding Your Starting Point

If you're new to IT, start with foundational certifications to build the essential knowledge of how computers and networks work. CompTIA A+ provides basic computer literacy and fundamental IT concepts, while ISC2 CC introduces basic security concepts. Either can serve as a starting point: Choose A+ if you need general IT foundations, or CC if you want to begin directly with security fundamentals. Network+ is particularly valuable because it teaches networking fundamentals essential for security work. You cannot effectively secure a network until you understand how it functions. Security+ is your next logical step. It's your gateway to the field and is widely recognized across the industry.

If you have IT experience, begin directly with Security+. It covers core security concepts and appears frequently in entry-level security job postings. Your existing IT knowledge provides the foundation that Security+ builds upon.

Choosing Your Specialization Path

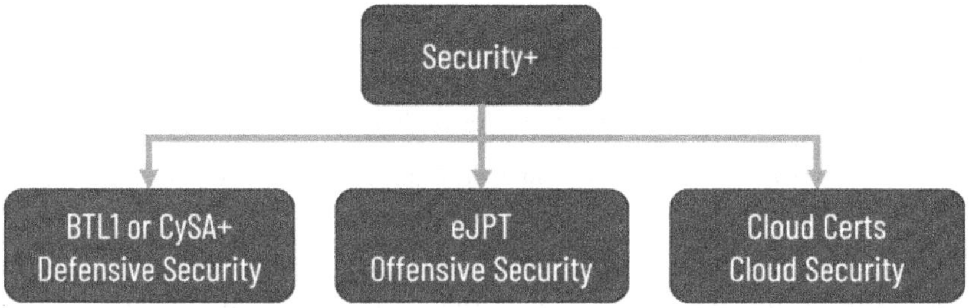

Figure 12.2: Choosing Your Specialization Path

Choosing Your Specialization

Once you've earned Security+ (or if you already have it), you can then pursue specialized certifications based on your career interests:

Defensive Security (BTL1 or CySA+): If you are interested in protecting systems and responding to threats, both BTL1 and CySA+ focus on defensive security roles like SOC analyst and incident responder. Choose BTL1 for hands-on lab experience with security analysis tasks, or choose CySA+ for analytical skills in threat detection and response. You can pursue either BTL1 or CySA+ alongside or after Security+. Choose based on your learning style and whether you prefer hands-on labs (BTL1) or analytical focus (CySA+).

Offensive Security (eJPT): If you're drawn to ethical hacking and penetration testing, the eJPT demonstrates your ability to perform actual penetration testing tasks in a lab environment. I recommend building the foundation with Security+ first, as understanding defenses makes you a better penetration tester.

Cloud Security (Cloud Certs): If you are interested in cloud security roles, build a foundation with Security+ to understand general principles. Then, add cloud-specific certifications (Azure, AWS, or GCP) to specialize in securing cloud environments. As organizations flock to the cloud, cloud security skills are in high demand.

You don't need all of these specializations. Start with Security+ and one specialization based on your current interests. You can always add others as your career develops.

Gaining Practical Experience

Certifications demonstrate knowledge, but employers also want to see that you can apply that knowledge practically. Here's how you can gain hands-on experience even before landing your first cybersecurity job:

Online Training Platforms

Several platforms offer hands-on cybersecurity training through interactive labs and real-world scenarios. Because each platform has its own teaching style, I recommend exploring multiple options to find the one that resonates with you. Just as you have favorite actors, authors, or teachers, you will likely connect more strongly with certain instructors and platforms. There is no single "best" platform. The best one for you is simply the one that keeps you engaged and helps you learn effectively.

YouTube is an often-overlooked but incredibly valuable resource. Countless channels offer free tutorials covering everything from networking basics to advanced exploitation techniques. It is perfect for supplementing your learning and seeing concepts demonstrated in real time. You can watch experienced professionals work through challenges and demonstrate tools. Look for channels with consistent posting schedules and strong community engagement. These are usually reliable indicators of quality.

TryHackMe is one of the most beginner-friendly platforms. It offers guided learning paths, from absolute beginner to advanced, with hands-on labs for every concept. The platform gamifies learning, making it engaging and fun. Many paths are free, while a premium subscription unlocks additional content. TryHackMe also offers certifications for those who want to formally validate their skills.

HackTheBox is more challenging and geared toward those with foundational knowledge. It offers realistic penetration testing scenarios where you hack into vulnerable machines to capture flags. While it offers certifications like TryHackMe, its primary value lies in the hands-on learning experience.

LetsDefend focuses specifically on Blue Team and defensive security skills. The platform provides realistic SOC analyst training through simulated incidents and alert investigations. It is an excellent choice for aspiring SOC analysts or incident responders, as it provides practical experience with the tools and workflows needed for these positions.

Cybrary provides a wide range of courses, from beginner fundamentals to advanced specializations. While course quality can vary, it is a good resource for exploring different areas of cybersecurity to find what interests you most.

Udemy is a massive marketplace featuring hundreds of cybersecurity instructors—including me. Because it hosts some of the biggest names in the industry, you are bound to find a teaching style that resonates with you. However, a tip: Rarely pay full price, as courses frequently go on sale. Since quality varies, check reviews and prioritize courses that include hands-on labs.

INE (formerly eLearnSecurity) offers comprehensive training covering everything from networking fundamentals to advanced penetration testing. Their training is known for being high-quality and practical, with hands-on labs integrated throughout.

Capture the Flag (CTF) Competitions

Capture the Flag (CTF) competitions are cybersecurity challenges where you solve puzzles, exploit vulnerabilities, and capture "flags" to earn points. They're excellent for developing practical skills and learning to think like both an attacker and a defender. Many CTFs are beginner-friendly and provide learning resources. Websites like CTFtime.org list upcoming competitions.

Home Labs

Building your own home lab is one of the best ways to gain hands-on experience. A home lab allows you to experiment freely, break things without consequences, and learn by doing. You can set up vulnerable machines, practice attacks and defenses, configure security tools, and simulate real-world scenarios.

While you could buy dedicated physical servers, virtualization software is the most economical way to build a powerful lab. This software lets you run multiple virtual computers, each with its own operating system and tools, right on your existing desktop or laptop, all in a self-contained environment.

Here are some of the most popular options to get you started:

- **VirtualBox**: Free and open-source virtualization software that runs on Windows, Mac, and Linux. Great for beginners.
- **VMware Workstation Player**: Free for personal use (Pro version is paid). Offers excellent performance and features.
- **Proxmox:** A powerful open-source virtualization platform that runs on dedicated hardware. Excellent for more advanced home labs with multiple virtual machines and networks. Proxmox is particularly popular among enthusiasts because it's free and offers enterprise-level features.

Once you have virtualization set up, you can download vulnerable machines from sites like VulnHub or use intentionally vulnerable VMs like Metasploitable to practice your skills. You can also set up your own networks, configure firewalls, deploy security tools like SIEM systems, and experiment with different configurations.

> **KEY POINT: Practicing on Systems You Don't Own**
>
> Practicing attacks on systems you don't own or without explicit written permission is illegal, even if your intentions are purely educational. This is a critical point that many beginners don't fully understand. Unauthorized access to computer systems is a federal crime in most countries, and "I was just learning" isn't a legal defense.

Only Practice On:

- Your own home lab virtual machines
- Vulnerable machines designed for practice (VulnHub, Metasploitable, DVWA)
- Authorized CTF platforms (TryHackMe, HackTheBox, CTF competitions)
- Systems with explicit written permission

Never Practice On:

- Production systems without written authorization—even at your own company
- Any system you don't own, regardless of assumptions about permission
- Public websites or services, even with good intentions
- School or university networks without authorized security research approval

Remember: Good intentions don't make unauthorized access legal. Many aspiring cybersecurity professionals have faced serious legal consequences, including criminal charges, because they didn't understand this boundary. Always get permission first, and always get that permission in writing.

Open Source Projects and GitHub

Contributing to open-source security projects or creating your own security tools demonstrates practical skills and gives you something tangible to show employers. Even simple projects like security scripts, automation tools, or documentation contributions can strengthen your resume and GitHub profile and demonstrate your ability to work with code (a skill many security roles require).

Bug Bounty Programs

Once you've developed some skills, consider participating in bug bounty programs where organizations pay security researchers to find and report vulnerabilities in their systems. Platforms like HackerOne and Bugcrowd host bug bounty programs for major companies. This provides real-world experience and can potentially generate income. However, be aware that bug hunting is highly competitive and often requires a significant time investment before you see a payout.

Volunteer and Pro Bono Work

Consider offering your cybersecurity skills to non-profit organizations, small businesses, or community groups. Many small organizations need help with basic security, but can't afford professional services. Start with clearly defined projects like security assessments, policy development, or employee training. Don't take on full security program responsibility immediately. Volunteering provides real-world experience, builds your portfolio, and helps your community. Even for volunteer work, document what you're providing in a simple written agreement to protect both you and the organization.

Building Your Professional Network

Networking is crucial in cybersecurity. Many jobs are filled through referrals and personal connections before they're ever publicly posted. Here's how to build your network:

- **Attend local cybersecurity meetups and conferences** (both in-person and virtual)—look for BSides conferences, OWASP chapters, and ISC2 chapter meetings.
- **Join online communities** like Reddit's r/cybersecurity, Discord servers, and professional forums.
- **Participate in LinkedIn discussions** and connect with cybersecurity professionals.
- **Engage with the cybersecurity community on Twitter/X** where many professionals share knowledge and opportunities.
- **Join professional organizations** like ISC2, ISACA, or local chapters of security groups.

Don't be afraid to reach out to people for informational interviews or mentorship. Most cybersecurity professionals are happy to help newcomers because they remember what it was like when they were starting out.

Crafting Your Path Forward

There's no single "right" path into cybersecurity. Your journey will depend on your background, interests, learning style, and circumstances. However, here's a general roadmap that works for many people:

1. **Build foundational IT knowledge** (if you don't already have it) through self-study, certifications like A+ and Network+, or entry-level IT work experience.
2. **Pursue an entry-level security certification** like ISC2 CC or CompTIA Security+ to validate your security knowledge.
3. **Gain hands-on experience** through online platforms like TryHackMe and HackTheBox, building a home lab, and participating in CTFs.
4. **Specialize in an area that interests you** by pursuing additional certifications (CySA+, BTL1, eJPT, cloud security, etc.) and focused training.
5. **Build a portfolio** of projects, write-ups, and contributions that demonstrate your practical skills.
6. **Network actively** and apply for entry-level positions, internships, or junior roles.
7. **Continue learning and growing** throughout your career, staying current with new threats, technologies, and techniques.

Remember: Everyone's timeline is different. Some people break into cybersecurity within months; for others, it takes years. Don't rush the process—give yourself the time you need to truly learn. What's most important is that you truly understand the concepts you're learning, not just memorize them for an exam. Focus on building deep comprehension rather than surface-level knowledge, as that understanding will serve you throughout your career. Go at your own pace. Your journey is unique to you, and comparing yourself to others will only create unnecessary stress. What matters is consistent progress and continuous learning, not how quickly you reach each milestone.

Bringing It All Together

You now have the foundational knowledge of cybersecurity and a practical roadmap for breaking into the field. This chapter has shown you that cybersecurity offers diverse career paths, including defensive blue team roles, offensive red team positions, technical security engineering, and governance and compliance work.

The path forward requires three key elements working together: knowledge, practical skills, and professional connections. Certifications like Security+, BTL1, and eJPT validate your knowledge and help you stand out to employers. Hands-on experience on platforms like TryHackMe, HackTheBox, and home labs demonstrates that you can apply that knowledge in practice. And networking within the cybersecurity community opens doors to opportunities that aren't publicly advertised.

The reality is that breaking into cybersecurity takes dedication and patience. Most people benefit from foundational IT experience before transitioning to security roles. However, direct entry is possible with focused self-study and demonstrated practical skills. Entry-level positions like Security Analyst, Junior Incident Responder, and Security Operations Specialist offer competitive salaries and strong growth potential for anyone willing to invest in their development.

What matters most is consistent progress and continuous learning. Cybersecurity is a field that rewards curiosity, persistence, and the willingness to constantly adapt to new threats and technologies. The global shortage of qualified cybersecurity professionals means there's never been a better time to enter this field. Organizations desperately need talented people who understand security fundamentals and can apply them practically.

Your cybersecurity journey starts with taking the first step, whether that's pursuing your first certification, building your first home lab, completing your first TryHackMe room, or applying for your first security role. Where that journey takes you depends on your dedication, curiosity, and willingness to never stop learning.

Key Takeaways

- **Cybersecurity offers diverse career paths**. It's more than hacking. Explore defensive, compliance, and engineering roles before specializing.
- **Build your IT foundation**. Most security roles build on IT fundamentals; start with A+ and Network+ if you're new to the field.
- **Certifications validate your knowledge**. Credentials like Security+, CC, and BTL1 prove your skills to employers.
- **Start with affordable certifications**. Avoid expensive training until you're employed; focus on recognized, foundational credentials first.
- **Hands-on experience is essential**; use platforms like TryHackMe, HackTheBox, and home labs to build practical skills.
- **Practice ethically and legally**: never test systems without explicit written permission; unauthorized hacking is a crime.

- **Choose learning resources that work for you**—whether it's hands-on labs, video courses, or reading, engagement matters more than any specific platform.
- **Network to find hidden opportunities**; many jobs are filled through referrals before they're ever posted publicly.
- **Learn at your own pace**. Focus on deep understanding, not speed; your journey is unique to you.

Quick Action: Take Your First Career Step

This chapter provided your roadmap for breaking into cybersecurity, but reading about career paths and certifications doesn't move you forward. Action does. The biggest barrier most people face isn't lack of knowledge or ability; it's simply not taking the first step. This quick action gets you started right now, transforming "I want to work in cybersecurity someday" into "I'm actively building cybersecurity skills today."

Begin Your Learning Journey Right Now

Step 1: Create Your First Hands-on Learning Account

- Choose one platform based on your interest:
  - **TryHackMe (tryhackme.com)**: Beginner-friendly with guided, gamified learning paths
  - **HackTheBox (hackthebox.com)**: More challenging, focuses on offensive security skills
  - **LetsDefend (letsdefend.io)**: Focuses on defensive security and SOC analyst skills
- Create a free account and browse the available tutorials and learning paths.
- Don't worry about which platform is "best." They're all excellent, just pick one and start.

Step 2: Complete One Beginner Challenge

- Pick any challenge marked "beginner," "tutorial," or "easy" (most take 15-30 minutes) and follow the instructions to work through it.
- Don't worry about getting it perfect. The goal is to start. If you get stuck, that's normal. Just check the hints or community forums.
- Completing even a simple challenge proves to yourself: "I can do this!"

Congratulations! You've just done hands-on cybersecurity work, which is something most people only talk about. You're no longer someone who's "interested in cybersecurity." You're someone who's actively building cybersecurity skills, and tomorrow you can come back and complete another challenge to keep that momentum going, because consistent progress is what builds careers.

Chapter 13. Conclusion & Congratulations

Congratulations

Congratulations on completing this journey through *Cybersecurity Fundamentals for Beginners*. We covered a massive amount of ground together, from the CIA Triad and networking fundamentals to online threats and the digital protection that stops them. You've navigated the technical, operational, and managerial sides of security. You now have a comprehensive understanding of the entire cybersecurity landscape, and that is a tremendous achievement. You're no longer an absolute beginner. You now have the strong foundational understanding of cybersecurity that I set out to provide. This serves you well whether you are pursuing a career in the field or simply want to be more informed in your personal and professional life.

Moving forward, I encourage you to keep the momentum going. Go back and review any concepts that you feel need a second look. Your goal now is to turn knowledge into security action. The skills you've learned are in high demand, and the world desperately needs your dedication. Continue practicing, seeking out certifications, and gaining hands-on experience. Remember: This book was your starting line, and your exciting cybersecurity journey begins right now. I'm confident you have everything you need to succeed.

Don't Forget Your Cybersecurity Course

As a thank you, everyone who purchased this book receives free enrollment into my *Cybersecurity for Absolute Beginners: Essential Concepts* course at InstructorAlton.com, which normally sells for $49. The online course features video lectures, live demonstrations, and additional resources that complement the material in this book. To access the course, please provide proof of purchase by following the steps below:

1. Take a screenshot or picture of your receipt as proof of purchase
2. Email your proof of purchase to alton@instructoralton.com
3. I'll email you your free enrollment code

Discounted & Free IT Courses

As a special thank you to our customers, we'd like to take this opportunity to offer you all of our IT courses for **50% OFF**!

Visit **InstructorAlton.com** and use the following coupon code when enrolling: **Amazon50OFF**

One Last Thing...

If you enjoyed this book and found it useful, I'd be very grateful if you'd post a short review on Amazon. Your support really does make a difference. I read all the reviews personally to get your feedback and make this book even better.

Thanks again for your support!

Glossary of Terms

A

Access Control: Security measures that regulate which people or systems can access or use resources in a computing environment. This includes authentication (verifying identity) and authorization (granting permissions).

Accounting (or Auditing): The process of tracking user activities, system resource usage, and network access to create an auditable record of who did what, when, and how. It is the third component of the AAA security framework.

Adware: Unwanted software designed to show advertisements or collect marketing-type data about users, often installed without consent and capable of acting as spyware.

AI-Enhanced Malware: Malware that uses AI (artificial intelligence) to improve its ability to evade detection, adapt to security measures, and personalize attacks to increase success rates.

ARP (Address Resolution Protocol): A protocol used to map IP addresses to MAC addresses on a local network; vulnerable to spoofing attacks.

Attack Surface: The total number of access points an unauthorized user can use to try to enter data into or extract data from a network or system. The larger the attack surface, the more vulnerable an organization is to cyberattacks.

Authentication: The process of verifying the identity of a user, device, or system. Common authentication methods include passwords, biometrics, security tokens, and multi-factor authentication.

Authorization: The security process that determines what an authenticated user or system is permitted to access, view, or perform after they have successfully proven their identity. It is the second component of the AAA security framework.

B

Backdoor: A method of bypassing normal authentication or security controls to gain access to a system. Attackers install backdoors to maintain persistent access even after their initial entry point is discovered.

Black Box Testing: A penetration testing approach where the ethical hacker has no prior knowledge of the target systems, simulating an external attacker's perspective.

Black Hat Hacker: A hacker who violates computer security for malicious reasons, personal gain, or both. They exploit vulnerabilities without permission and often with criminal intent.

Botnet: A network of compromised computers controlled by an attacker, often used to launch distributed denial-of-service (DDoS) attacks or send spam.

Brute Force Attack: An attack method that tries every possible combination of characters to crack a password or encryption key.

Buffer Overflow: A vulnerability that occurs when more data is written to a memory buffer than it can hold, potentially allowing attackers to execute malicious code.

Business Continuity: An organization's ability to maintain essential functions during and after a disaster or disruption. Business continuity planning ensures that critical operations can continue with minimal interruption.

Business Email Compromise (BEC): A sophisticated scam targeting businesses where attackers impersonate executives or trusted partners to trick employees into transferring money or revealing sensitive information.

C

CIA Triad: The three fundamental principles of information security: Confidentiality (preventing unauthorized disclosure), Integrity (preventing unauthorized modification), and Availability (ensuring systems and data are accessible when needed).

Cloud Computing: The delivery of computing services—including servers, storage, databases, networking, software, and analytics—over the Internet ("the cloud") rather than from local servers or personal computers.

Compliance: Adherence to laws, regulations, standards, and policies that apply to an organization. In cybersecurity, this often includes industry-specific regulations like HIPAA for healthcare or PCI DSS for payment card processing.

Critical Infrastructure: Essential systems and assets vital to a nation's security, economy, and public health or safety. Examples include power plants, water systems, hospitals, and financial institutions.

Cryptography: The practice of developing and analyzing encryption algorithms and security protocols to protect sensitive information.

Cyber Terrorist: An individual or group that uses hacking and cyberattacks to intimidate or coerce a government or civilian population, typically in pursuit of political or ideological goals.

D

DAD Triad: A conceptual framework representing the three main security threats: Disclosure (confidentiality loss), Alteration (integrity loss), and Destruction/Denial (availability loss). These threats work against the goals of the CIA Triad.

Dark Web: A part of the Internet that is intentionally hidden and requires specific software (like Tor) to access. It's often associated with illegal activities, though it also has legitimate uses for privacy and anonymity.

Data Breach: An incident where sensitive, protected, or confidential data is accessed, stolen, or used by an unauthorized individual. This can include personal information, financial data, intellectual property, or trade secrets.

DDoS (Distributed Denial of Service): A cyberattack that uses multiple compromised systems to overwhelm a target with traffic, making services unavailable to legitimate users.

Deepfake: AI-generated synthetic media (audio, video, or images) that realistically depicts people saying or doing things they never actually said or did; used for fraud and misinformation.

Defense in Depth: A layered approach to IT security where multiple different types of security controls are implemented throughout an organization, network, and IT systems to significantly reduce risks and vulnerabilities.

DHCP (Dynamic Host Configuration Protocol): A protocol that automatically assigns IP addresses to devices when they connect to a network.

DMZ (Demilitarized Zone): A perimeter network designed to be securely separated from an organization's internal private network; used to host public-facing services while protecting the internal network.

DNS (Domain Name System): A protocol that translates human-readable domain names (like www.example.com) into IP addresses that computers use to identify each other on the network.

DoS (Denial of Service): A cyberattack that overwhelms a system with requests from a single source, making services unavailable to legitimate users.

Drive-By Download: A method of deploying malware where visiting an infected website automatically downloads and installs malware without the user's knowledge or consent.

Dumpster Diving: A physical security threat where attackers search through trash and dumpsters for sensitive information that can be used for malicious purposes.

Dwell Time: The length of time that an attacker is able to access a network before they are detected and removed. Shorter dwell times generally result in less damage and lower costs.

E

Encryption: The process of converting information or data into a code to prevent unauthorized access. Encrypted data can only be read by someone who has the key to decode it.

Endpoint Protection: Security measures implemented on individual devices (endpoints) such as computers, servers, and mobile devices to protect them from threats. Includes anti-malware software, firewalls, and system hardening.

Ethical Hacker: See White Hat Hacker.

Exploit: ***(n)*** A piece of software, a sequence of commands, or a technique that takes advantage of a vulnerability to gain unauthorized access or cause unintended behavior in a system; ***(v)*** To take advantage of a vulnerability to gain unauthorized access to a system in order to cause unintended behavior.

F

File Hash: A unique fixed-length string of characters generated by a cryptographic algorithm (such as MD5, SHA-1, or SHA-256) that serves as a digital fingerprint for a file. File hashes are used to verify file integrity, detect malware, and confirm that files haven't been altered or corrupted.

Firewall: A network security device or software that monitors and controls incoming and outgoing network traffic based on predetermined security rules; a barrier between a trusted internal network and untrusted external networks.

G

GDPR (General Data Protection Regulation): A comprehensive data protection law in the European Union that regulates how organizations collect, store, and use personal data. It includes strict requirements and significant penalties for non-compliance.

Gray Hat Hacker: A hacker who operates in a moral and legal gray area, hacking into systems without permission but with the intention of finding and reporting vulnerabilities rather than exploiting them for personal gain or malicious reasons.

Grey Box Testing: A penetration testing approach where the ethical hacker has limited knowledge and access to the target systems, simulating an insider threat or an attacker who has already gained initial access.

H

Hacktivist: A hacker who uses their skills to promote a political agenda or social change. They typically target organizations they view as opposing their cause.

Hacking: The act of gaining unauthorized access to computer systems, networks, or data. While the term can technically apply to any clever technical solution, in cybersecurity, it typically refers to exploiting vulnerabilities to bypass security measures.

Hacking as a Service (HaaS): A business model where cybercriminals offer hacking services for hire, allowing less technically skilled individuals to conduct sophisticated cyberattacks for a fee.

Honeypot: A decoy system designed to lure attackers away from systems that actually contain sensitive data. It distracts attackers, allowing security teams to study attack methods and gather intelligence.

I

IDS (Intrusion Detection System): A device or software application that monitors a network or systems for malicious activity or policy violations. When detected, incidents are typically reported to a security operations center. IDS responds passively by logging and alerting.

Impersonation: A social engineering tactic where an attacker pretends to be someone they're not (such as an IT technician, bank representative, or coworker) to gain trust and manipulate victims.

Incident Response: The process of preparing for, detecting, analyzing, and recovering from security incidents. Incident response professionals work to contain damage, investigate what happened, and restore normal operations.

IoT (Internet of Things): A network of physical devices embedded with sensors, software, and connectivity that allows them to collect and exchange data over the Internet. Examples include smart thermostats, fitness trackers, and connected home security systems.

IP Address (Internet Protocol Address): A logical address assigned to a device on a network; used for routing data across networks and the Internet.

IPS (Intrusion Prevention System): A network security device that monitors for malicious activity and responds actively by changing network configurations to stop attacks, such as modifying access control lists or blocking ports.

IPv4: The fourth version of the Internet Protocol, using 32-bit addresses displayed as four decimal numbers (e.g., 192.168.1.1).

IPv6: The sixth version of the Internet Protocol, using 128-bit addresses to provide a vastly larger address space than IPv4.

L

LAN (Local Area Network): A computer network within a small geographical area, such as a single room, building, or group of buildings, where all devices are directly connected.

Least Privilege: A security principle stating that a user, system, process, or application should only be given the permissions necessary to complete its assigned tasks or functions and nothing more.

Logic Bomb: A string of malicious code embedded into an application designed to execute a malicious payload in response to a specific event or condition, such as a particular date, time, or a specific action being taken (e.g., removal of an employee's access).

M

MAC Address (Media Access Control Address): A unique physical address burned into a network interface card (NIC) that identifies a device on a local network.

Malicious Hacker: See Black Hat Hacker.

Malware: Malicious software designed to intentionally damage, destroy, steal data, or gain unauthorized access to a computer system.

Multi-Factor Authentication (MFA): A layered approach to authentication that requires two or more verification factors: something you know, something you have, and/or something you are.

N

Nation-State Actors: Hackers or hacking groups that are sponsored or directed by a government to conduct cyber operations against other nations, organizations, or individuals for political, military, or economic purposes.

Network Fingerprinting: The process of identifying the operating systems, services, and configurations of devices on a network to understand the network's architecture and identify potential vulnerabilities.

Non-Repudiation: A core security principle that ensures a party cannot successfully deny having performed a specific action, such as sending a message or executing a transaction. It provides verifiable proof of origin and integrity.

O

OSINT (Open-Source Intelligence): Information collected from publicly available sources such as websites, social media, public records, and press releases; used in the information gathering phase of hacking.

P

Passphrase: A sequence of random words strung together to create a memorable but secure password, typically longer than traditional passwords (15–20+ characters vs. 8–12 characters).

Password Manager: Software that generates, stores, and manages strong, unique passwords for multiple accounts, requiring users to remember only one master password.

Penetration Test: A simulated cyberattack against a computer system, network, or application to identify vulnerabilities that could be exploited by malicious hackers. Also called a pen test.

Penetration Test Report: A comprehensive document produced by ethical hackers detailing discovered vulnerabilities, successful exploits, potential business impacts, and prioritized recommendations for remediation.

Phishing: A social engineering cyberattack that uses fraudulent emails, text messages, or websites to trick people into revealing sensitive information, downloading malware, or taking actions that compromise security.

Piggybacking (Tailgating): A physical security threat where an unauthorized person gains access to a restricted area by following closely behind an authorized person (e.g., through a door opened by the authorized person).

Pre-Engagement: The phase before a penetration test begins, where the ethical hacker and client define the scope, rules of engagement, objectives, and legal boundaries of the security assessment.

Pretexting: A social engineering technique where an attacker creates a fabricated scenario (pretext) to manipulate victims into divulging information or performing actions.

Privilege Escalation: The act of exploiting a vulnerability or design flaw to gain elevated access to resources that are normally protected, such as gaining administrator rights from a standard user account.

Q

Quishing (QR Code Phishing): An emerging phishing threat where attackers use QR codes in emails, text messages, or physical posters to redirect victims to malicious websites.

R

Ransomware: A type of malicious software that encrypts a victim's files or locks them out of their systems. The attacker then demands a ransom payment in exchange for restoring access.

Ransomware as a Service (RaaS): A business model where cybercriminals sell ransomware, allowing less technically skilled individuals to conduct sophisticated cyberattacks for a fee.

Risk: The potential for loss, damage, or destruction of an asset when a threat exploits a vulnerability.

Risk Assessment: The process of identifying and evaluating IT risks by analyzing assets, threats, vulnerabilities, probability, and impact to prioritize security responses.

Risk Management: The process of identifying, assessing, monitoring, and limiting risk to an acceptable level through avoidance, acceptance, mitigation, or transference.

Rootkit: Malware designed to gain root (administrative) access to a system by exploiting vulnerabilities that enable privilege escalation. Rootkits modify core system files and are often invisible to the operating system and anti-malware programs.

Router: A network device that connects multiple networks together and uses IP addresses and routing protocols to forward data between networks.

Routing Protocol: A protocol that enables routers to communicate with each other, share network topology information, and determine the best paths for data transmission.

S

Script Kiddie: An unskilled individual who uses existing computer scripts or programs to hack into systems, often without understanding how they work. The term is generally derogatory within the hacking community.

Smishing (SMS Phishing): Phishing attacks conducted via fraudulent text messages to trick victims into clicking malicious links or providing personal information.

Social Engineering: The psychological manipulation of people into performing actions or divulging confidential information. It exploits human nature rather than technical vulnerabilities.

Spear Phishing: A targeted phishing attack aimed at specific individuals or organizations, using personalized information gathered from research to make the attack more convincing.

Spoofing: The malicious forging of email addresses, communication headers/logos, or identities to make messages appear to come from a trusted source when they actually come from an attacker.

Spyware: Any form of malware designed to spy on users without their consent or knowledge, collecting information such as keystrokes, screenshots, authentication credentials, or personally identifiable information.

SQL Injection: A web-based attack that exploits poor input validation to execute malicious database commands, potentially allowing attackers to access, modify, or delete database contents.

State-Sponsored Hacker: A hacker who conducts cyber operations on behalf of a nation-state, typically targeting other governments, critical infrastructure, or organizations of strategic interest.

Supply Chain Attack: A cyberattack that targets a less-secure element in the supply chain to compromise a final target. Attackers infiltrate a trusted vendor or partner to gain access to their customers' systems.

Switch: A network device that connects multiple LAN devices and uses MAC addresses to intelligently forward data to the correct destination device.

Symmetric Encryption: An encryption method that uses a single key for both encryption and decryption, also known as private key encryption.

T

Tailgating: See Piggybacking.

TCP (Transmission Control Protocol): A connection-oriented transport protocol that ensures reliable, ordered, and error-checked delivery of data.

TCP/IP Protocol Suite: A set of networking protocols organized into four layers (Application, Transport, Internet, and Network Access) that enables communication across networks and the Internet.

Threat Actor: A generic term for any individual, group, or organization (e.g., hacker, criminal syndicate, nation-state, hacktivist) that poses a threat to information security. Also known as a malicious actor.

Traceroute: A network diagnostic tool that displays the route and measures transit delays of packets across a network, showing each router hop along the path.

Trojan Horse: Malware disguised as desirable software that includes hidden malicious code, deceiving users into installing malware by appearing legitimate.

U

UDP (User Datagram Protocol): A connectionless transport protocol that provides fast but unreliable data transmission without error checking or guaranteed delivery.

V

Vishing (Voice Phishing): Phishing attacks conducted via phone calls where attackers impersonate trusted organizations to trick victims into revealing sensitive information.

VPN (Virtual Private Network): A technology that creates a secure, encrypted connection over a less secure network, such as the Internet. VPNs allow remote workers to securely access their organization's network from anywhere.

Vulnerability: A weakness in a system, application, or network that could be exploited to compromise security. Vulnerabilities can exist in software, hardware, configurations, or even human activity.

Vulnerability Assessment: The process of identifying, evaluating, and prioritizing security vulnerabilities in systems and applications.

W

WAN (Wide Area Network): A computer network that extends over a large geographical distance, typically connecting multiple LANs across cities and countries.

Whaling: A type of spear phishing that specifically targets high-profile individuals such as CEOs, CFOs, or other senior executives.

White Box Testing: A penetration testing approach where the ethical hacker has complete knowledge of the target systems, including architecture diagrams, source code, and credentials, allowing for a comprehensive security assessment.

White Hat Hacker (Ethical Hacker): A cybersecurity professional who uses hacking techniques to identify, report, and/or fix security vulnerabilities. They work with permission and within legal boundaries to improve security.

Wireless Security: Security measures implemented to protect wireless networks, including encryption protocols (WPA2/WPA3), strong passwords, and authentication mechanisms.

Worm: Malicious software that can travel and replicate autonomously throughout a network without user interaction or host application assistance, often consuming network bandwidth and degrading performance.

X

XSS (Cross-Site Scripting): A web-based attack that injects malicious scripts into trusted websites, which then execute in users' browsers, potentially stealing session cookies or credentials.

Z

Zero-Day Attack: A cyberattack that exploits a previously unknown vulnerability in software or hardware before the vendor has released a patch or fix.

Made in the USA
Monee, IL
14 June 2026

54109977R10168